T0225894

Snowflake Access Control

Mastering the Features for Data Privacy and Regulatory Compliance

Jessica Megan Larson

Snowflake Access Control: Mastering the Features for Data Privacy and Regulatory Compliance

Jessica Megan Larson
Oakland, CA, USA

ISBN-13 (pbk): 978-1-4842-8037-9   ISBN-13 (electronic): 978-1-4842-8038-6
https://doi.org/10.1007/978-1-4842-8038-6

Copyright © 2022 by Jessica Megan Larson

This work is subject to copyright. All rights are reserved by the Publisher, whether the whole or part of the material is concerned, specifically the rights of translation, reprinting, reuse of illustrations, recitation, broadcasting, reproduction on microfilms or in any other physical way, and transmission or information storage and retrieval, electronic adaptation, computer software, or by similar or dissimilar methodology now known or hereafter developed.

Trademarked names, logos, and images may appear in this book. Rather than use a trademark symbol with every occurrence of a trademarked name, logo, or image we use the names, logos, and images only in an editorial fashion and to the benefit of the trademark owner, with no intention of infringement of the trademark.

The use in this publication of trade names, trademarks, service marks, and similar terms, even if they are not identified as such, is not to be taken as an expression of opinion as to whether or not they are subject to proprietary rights.

While the advice and information in this book are believed to be true and accurate at the date of publication, neither the authors nor the editors nor the publisher can accept any legal responsibility for any errors or omissions that may be made. The publisher makes no warranty, express or implied, with respect to the material contained herein.

Managing Director, Apress Media LLC: Welmoed Spahr
Acquisitions Editor: Jonathan Gennick
Development Editor: Laura Berendson
Coordinating Editor: Jill Balzano

Cover designed by eStudioCalamar

Cover image designed by Freepik (www.freepik.com)

Distributed to the book trade worldwide by Springer Science+Business Media LLC, 1 New York Plaza, Suite 4600, New York, NY 10004. Phone 1-800-SPRINGER, fax (201) 348-4505, e-mail orders-ny@springer-sbm.com, or visit www.springeronline.com. Apress Media, LLC is a California LLC and the sole member (owner) is Springer Science + Business Media Finance Inc (SSBM Finance Inc). SSBM Finance Inc is a **Delaware** corporation.

For information on translations, please e-mail booktranslations@springernature.com; for reprint, paperback, or audio rights, please e-mail bookpermissions@springernature.com.

Apress titles may be purchased in bulk for academic, corporate, or promotional use. eBook versions and licenses are also available for most titles. For more information, reference our Print and eBook Bulk Sales web page at http://www.apress.com/bulk-sales.

Any source code or other supplementary material referenced by the author in this book is available to readers on GitHub at https://github.com/Apress/snowflake-access-control/releases/tag/v1.0.

Printed on acid-free paper

I want to dedicate my book to all of my amazing friends, family, and colleagues that have continued to encourage me through this process; I could not have done it without all of you.

Table of Contents

About the Author

Jessica Megan Larson was born and raised in a small town across the Puget Sound from Seattle, but now calls Oakland, California, home. She studied cognitive science with a minor in computer science at the University of California Berkeley. She thrives on mentorship, solving data puzzles, and equipping colleagues with new technical skills. Jessica is passionate about helping women and non-binary people find their place in the technology industry. She was the first engineer within the Enterprise Data Warehouse team at Pinterest, and additionally helps to develop fantastic women through Built by Girls. Previously, she wrangled data at Eaze and Flexport. Outside of work, Jessica spends her time soaking up the California sun playing volleyball on the beach or at the park.

About the Technical Reviewer

 **Iliana Iankoulova** is an experienced professional in the field of data warehousing and for the past years has been solving data challenges at the fastest growing online supermarket in Europe: Picnic Technologies. She joined as the first data engineer, and is currently leading several R&D teams responsible for data governance, analytics, data warehousing in Snowflake, and master data management. Iliana started her career over a decade ago as a business intelligence consultant in banking and e-commerce. She holds an M.Sc. (cum laude) in Business Information Technology from Twente University, the Netherlands, and a B.Sc. (first-class honors) in Computing and Business from Brock University, Canada. She is a published author on the topics of cloud data security, business intelligence, and data engineering. She is passionate about Kimball's dimensional modeling and Linstedt's Data Vault. Iliana is Bulgarian-Canadian and naturally found the perfect home in Rotterdam, the Netherlands.

Acknowledgments

I want to thank my amazing editors, Jonathan Gennick, Jill Balzano, and Laura Berendson, who've fielded so many random questions throughout this process and have shown so much patience. I also want to thank my technical reviewer, Iliana Iankoulova, who has been outstanding in identifying ways I could improve this book delivered with kindness and candor.

Introduction

Traditionally, databases have provided few features and an environment focused mainly on efficiently storing data in a way that can be fetched with high performance. These databases were not designed for user-friendliness – this was a time when if you wanted to work with the database, you had to know your way around the command line. Enter the new wave of SaaS databases with a breadth of features and user interfaces a non-technical user could interact with. Snowflake led the pack on this, and quickly became a database known for its performance, as well as a ton of baked-in features making access control easier. They had the largest software Initial Public Offering (IPO) in history.

At the same time, consumers started to become aware of the privacy implications around their personal data – data that is held across a multitude of companies and exchanged fairly freely between them. As a result, we started to see an outpouring of new and attempted privacy laws across the world. Europe led the way with their General Data Protection Regulation (GDPR), a comprehensive set of regulations that dictate, among other things, what data companies can collect, how they can collect it, how they must store and protect it, and the role of the consumer's consent in all of this.

How This Book Is Structured

This book was designed to cover everything a data engineer or administrator may need to know in order to implement any level of access control in their Snowflake database in order to comply with existing regulations and beyond. As such, the book begins with the fundamental concepts that will help guide the understanding in later chapters.

Chapters 1–4 cover the basic concepts of access control, including the types of data that may require access control, the laws and regulations at play, and the different types of privileges Snowflake provides.

Chapters 5–9 cover roles in Snowflake – the different types of roles from a theoretical perspective, how a user works with their roles, secondary roles, and how to build roles using role inheritance.

Chapters 10–15 are the meat of the book, these chapters are when everything comes into play. These chapters cover the actual act of implementing access control at the different levels in the Snowflake database hierarchy.

Chapters 16–18 are geared more toward the operational side of things. These chapters will help guide process rather than implementation, and cover how we can create separate development and production environments, how we manage connections with upstream and downstream services, and how we manage incoming access requests at scale.

PART I

Background

CHAPTER 1

What Is Access Control?

Access Control is a broad term we use to describe any way that we allocate or restrict the ability to read, modify, delete, or create resources within a system. These objects may be self-contained systems like a server or a database, but more commonly, they're components of one of those systems, like a table or a schema, or even a single column or row within one of those tables.

Most of us make access control decisions every single day. When we post a picture of our dinner on social media, we decide who should be able to see the post. Within a larger organization, and when we're posting that homemade pasta, there are a number of reasons why we might want to restrict access to data. There may be sensitive information that few people should be able to view. We may have told mom that we returned her pasta machine weeks ago when we've been secretly holding on to it thinking she won't notice. Reporting datasets might require that only a specific service account be able to modify important financial figures. It's also possible that we're concerned about storage costs and we don't want just anyone adding new tables. Whatever the reasoning, access control is a crucial component of any data system.

When we decide to implement access control, the first question we need to ask ourselves is *What assets are we protecting?* We already know the *why* – we need to maintain compliance – and we can't yet determine the *how*. It is important we ask ourselves *what* because when we understand *what* we're protecting, it informs how we make decisions about everything else. For example, if we want to restrict access to customer phone numbers to the customer support agent actively helping them, we immediately note the kind of sensitive data we are working with, and the granularity of access we're seeking. From there, we can look to the applicable regulations for guidance. For sensitive personal data like phone numbers, it is likely that we will be affected by the EU's General Data Protection Regulation (GDPR), the California Consumer Protection Act (CCPA), and others that require that this data is only accessible on an as-needed basis. In this instance, we're protecting the customer's personal information, specifically one data field within one table, and we want to make sure that only the specific agent

© Jessica Megan Larson 2022
J. M. Larson, *Snowflake Access Control*, https://doi.org/10.1007/978-1-4842-8038-6_1

helping them can view that information. From there we can get a better idea of our philosophy around restricting access to that dataset. Once we know the paradigm, we can determine how we want to allocate access to users. This continues to waterfall until we're done solving the problem at hand. Since we're approaching access control using a problem-solving mindset, we can navigate this landscape with more clarity, rather than trying to massage our solution until it fits the problem.

Access Control Paradigms

There are many different philosophies behind access control. Many of these are not applicable to Snowflake, so they aren't as valuable to discuss at this juncture. We're going to focus on a few broad categories that are relevant for Snowflake. In Figure 1-1, we can visualize a few of these categories.

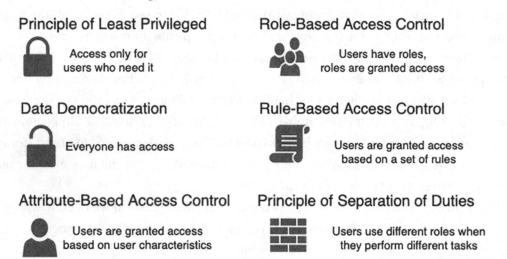

Figure 1-1. *A high-level overview of the access control paradigms*

All of these paradigms aim to answer the *why* but also the *who* and the *how* behind an access control system. Why are we allocating or restricting access to whom and how are we implementing it? In practice, access control is almost always an amalgam of multiple schools of thought.

Role-Based Access Control (RBAC)

Although phonetically similar to Row-Based Access Control, Role-Based Access Control assigns users access to data based on their role within an organization. When discussing with colleagues, to avoid confusion with row-based access control, I typically refer to row-based access control as row-level access control. Since groupings are created based on job description, engineers can see engineering data, sales can see sales data, and accounting can see financial data. One beautiful thing about RBAC is that segregating users into different groups based on their job function is often already done for other reasons. Accounting might have certain financial tools that other departments don't need to utilize, and as a result, many organizations already have a way to handle this. Importing those groupings into a database could be as simple as a short script that runs on a regular basis, keeping the database roles in sync with the rest of the organization.

Data Democratization

Data Democratization is a relatively new concept; it's rise facilitated by the explosion of big data. Organizations have never had so much data about their business, their users, and their industry, while at the same time, individuals have never had such an appetite to examine and disseminate the information at their disposal. In response, many organizations have made a push to make all or most data highly available to everyone within the organization to do with as they please. This can be a very beautiful thing, employees now have the power to examine the results of their labor and can use it to grow in their careers, make their team more efficient, or improve processes. However, this comes at the cost of privacy, oftentimes the consumer's.

Principle of Least Privilege (PLP)

Principle of Least Privilege requires that users only have access to information they need to fulfill their job functions. This is fundamentally the opposite of data democratization; however, both of these paradigms can thrive within an organization if boundaries are set accordingly. Principle of least privilege is the preferred and often-prescribed paradigm for working with highly sensitive datasets like personally identifiable information about users.

Rule-Based Access Control

Confusingly, this also abbreviates to RBAC, and if you're following – this is the third RBAC you have to worry about. For that reason, we are not going to refer to rule-based access control as RBAC in this book. Rule-based access control relies on a set of rules. These rules might restrict access based on a user's IP – for example, a rule could state that only users on the office virtual private network (VPN) can be allowed to log in to the database. These rules could be about anything from how a user is authenticating, including the infrastructure they're using, to attributes about the user themselves. This could also be used to restrict access to non-public material business metrics during a trading window to protect employees against insider trading.

Attribute-Based Access Control (ABAC)

Attribute-based access control secures data based on attributes of the user and attributes of the data. These attributes may include things like the region of a sales associate, the customer an agent has helped, or whether or not an individual is unwittingly victim to an elaborate theft of their pasta machine. These user attributes typically mirror the attributes of the data. Attribute-based access control lends itself to being combined with the other paradigms listed earlier as it is typically used for more granular access than others demand.

Principle of Separation of Duties

The Principle of Separation of Duties states that users should use different roles or accounts depending on the type of work they are doing. This means that a user should use an administrator role or administrator account when performing certain administrative actions like creating users or allocating access to data assets, and a separate account or role for viewing dashboards taking the pulse of the organization. This will come into play when we dive into Secondary Roles in a later chapter.

Access Control Methods

The different access control paradigms utilize some common concepts and rely on a few different methods of granting access to users. Since we want to allocate access at scale, it doesn't make sense to individually grant access to every dataset to every user. We can instead think about how we manage these grants in a few different methods. Just like the paradigms themselves, these methods can be used together or on their own.

Groups or Roles

We can assign each user to one or more groups or roles. We often think about groups and roles synonymously, but there is a distinction between the two. A group aims to capture a collection of users with commonalities. A role typically refers to the scope of work a user is doing. I also like to think about the directionality of groups vs. roles, users are typically added to a group, whereas a role is typically granted to a user. In Snowflake, we use roles, rather than groups, to define permissions for one or more users. These roles are then granted access to datasets. This drastically cuts down on the number of grants we need to execute and simplifies access greatly. In Snowflake, to grant access in the traditional way, we create the roles and in either order grant the roles to each user in a group and grant these roles access to datasets. This is the most commonly used method for role-based access control and the principle of separation of duties and can be used for the principle of least privileged as well as attribute-based access control.

Lookup Tables and Mappings

When we use lookup tables to manage access to assets, we are typically doing this for more granular access, such as column-level or row-level. Since this data lives in a table, it can be queried in SQL and joined to other tables to filter or mask data. We are limited by SQL's syntax and its ability to extract mapping information from a table, which means gating access to tables using this method is less ideal. A lookup table used to filter a sales leads dataset might have information like the sales regions different account managers serve. This is very commonly used for attribute-based access control and can be used for rule-based access control and the principle of least privileged. One downside to using lookup tables is that they will need to be continually updated and maintained as the organization grows and as the datasets evolve.

Miscellaneous Rules

There are also ways to limit access to data that don't fit neatly into the two preceding categories. The example about limiting viewing to a reporting dataset during trading windows doesn't involve roles or lookup tables but is still at its core access control. We will not spend much time on these ad hoc methods, as they deviate from typical access control patterns and are not scalable.

Wrapping Up

At a high level, access control seems like it is fundamentally a business problem rather than an engineering problem; however, everyone in an organization is responsible for safeguarding data. An organization decides how strictly they want to limit the access to data, while at the same time limiting the scope of data that can be exposed and used in inappropriate ways. Engineers and database administrators provide expertise and context for the organization, pointing out weak points where data could be exposed, and implementing the techniques outlined later in the book. As a data consumer, limiting the availability of data also limits an individual's liability. In this book, we will spend some time on the business aspects of access control, but our larger focus will be a deep dive into the engineering aspects of access control, homing in on tangible pieces an individual can perform.

CHAPTER 2

Data Types Requiring Access Control

Access control can be implemented across all types of data; however, not all types of data necessitate strict access control. Publicly available information like the GDP of the United States, the address of the closest coffee shop, and the winner of Album of The Year Grammy in 2021 are not of particularly sensitive nature. While those are not sensitive, I think we can all agree that Taylor Swift's cell phone number and credit card numbers are sensitive. The sensitivity of different types of data varies across the board.

Each type of data fits on a spectrum ranging from not sensitive to very sensitive, or even confidential, as shown in Figure 2-1.

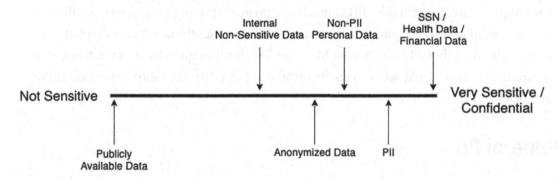

Figure 2-1. *Data ranges from not sensitive to very sensitive/confidential*

We can see in Figure 2-1 that our least sensitive information is publicly available data, and the most sensitive is personally identifiable information (PII), health data, and financial data. We want to safeguard different types of sensitive data for a multitude of reasons. We might want to protect the individuals because we have regulations requiring that we do so, like the European Union's General Data Protection Regulation (GDPR) or the California Consumer Privacy Act (CCPA) require, which we will dive into in the next chapter. We may want to mitigate liability, or we might do it because it's simply our

© Jessica Megan Larson 2022
J. M. Larson, *Snowflake Access Control*, https://doi.org/10.1007/978-1-4842-8038-6_2

organization's practice and we think it's the right thing to do. There's no right answer here as long as we are respecting the privacy of individuals, acting in an ethical manner, and complying with applicable regulations.

Before we start applying access control, we must identify the sensitivity of the data and then, combined with our access control philosophy, we can start to apply restrictions. This guarantees that we will be solving for the right problem and adequately protecting our sensitive data while at the same time not being overzealous with the solution we build. In this chapter, we will cover the different types of data and the spectrum of access control they demand. We're going to start with the most sensitive data on this spectrum, and we will move to the least sensitive types of data.

Personally Identifiable Information

Personally Identifiable Information (PII) is typically regarded as the most sensitive of commonly held data. PII is a broad term that encompasses all data that could be used to single out an individual, whether that individual is a consumer, an employee, a contractor, or any other person whose data may be held. Many different regulatory requirements center around PII and protecting consumers, which we will cover in the next chapter. As a general rule, PII should only be available to employees who absolutely need that data to do their job. However, not all PII is equal; there are certain pieces of information that shouldn't be accessible at all. We don't need to treat first names with the same care as we treat social security numbers. Let's dive into some of the different categories of PII.

General PII

We're using the term general to describe PII here because these are the types of PII that can apply to any of the following categories and are specific characteristics of individuals. These are the characteristics that come to mind if you were to describe a person: their name, how to contact them, and what they look like. Generally, these are characteristics that are unique data points that most individuals have. Some of the most common types of PII are

- First and last name
- Phone number

- Address

- Email

- Birthdate

- Social media account

- Driver's license number

- IP Addresses

There is some debate around whether or not first and last name belongs under PII because there could be more than one person sharing the same first and last name, and, therefore, the data alone does not isolate one individual. I prefer to err on the side of caution when a piece of data fits into two buckets by choosing the more sensitive bucket.

Multimedia

In addition to these personal characteristics, one's likeness can also be enough information to identify someone. Therefore, PII also includes multimedia, or references to multimedia, including the following:

- Photographs

- Video

- Audio

- GPS coordinates

- Cookies

All of the preceding are rich data types, that is, not nearly as simple as a six-digit birthdate or email address. Photographs and multimedia are typically stored in a cheaper way than a database, like in a service like Amazon S3 or Google Cloud Storage. Frequently, URLs linking to this data live in a database. Depending on how your organization handles these storage buckets, those URLs may need to be treated like PII as well.

Protected Health Information (PHI)

Protected Health Information (PHI) is a special category of PII due to the private nature of an individual's medical history and various regulations protecting that data. In the next chapter, we will explore regulations around working with health data. Examples of protected health information include the preceding PII categories combined with any of the following:

- Medical records numbers

- Test results

- Billing information

- Insurance account numbers

- Biometrics (iris, fingerprints, etc.,)

These types of data can be used to identify an individual in the case of biometrics and reports from medical professionals. Some of the data may not be able to pinpoint an individual, but are still considered PII because they are private health information that one should never need to disclose outside of a medical setting.

Financial Information

As you might imagine, with the complexity of our economy, there are many different types of sensitive financial data. In general, anything linked to a bank account, a card, a portfolio, or any other financial service is considered particularly sensitive and necessary to safeguard. Specifically, some of the data types you may see are

- Credit and debit card numbers

- Bank account numbers

- Credit scores

- Account usernames

- PINs and security codes

One or more of these pieces of information from an individual could be used to steal money from them, to create new accounts under their name, or take over assets.

Social Security Numbers and National Identity Numbers

Social security numbers (SSNs), or equivalent, are one of the most sensitive types of PII as they explicitly identify an individual; health records and financial information include SSNs as a particularly sensitive data type. While SSN is the United States' system for identifying individuals, the following are other equivalents around the world:

- European Social Security Number (ESSN)

- National identity number

- National identity document

- Social insurance number

- Central population registration

- Identity card

   In the United States, the SSN is a highly confidential number that an individual must safeguard. In other countries and economic unions, the identification numbers may not be as sensitive, and may be generated based on characteristics like first and last initial, birth month, and state of birth. In the example of the US SSN, this number is especially problematic because it can be used to steal an individual's identity and gain access to all of the other types of protected data. For this reason, we often refer to this type of data as confidential. Social Security Numbers can present a single point of failure in that an SSN alone is enough for a bad actor to inflict serious damage to an individual.

Passwords

Passwords are protected similarly to SSNs for much the same reason; a password combined with an email address, or an account ID, can be used by someone to gain access to all other forms of information and can potentially be used to steal an individual's identity. Passwords should never be stored as plain text and should never be introduced into an analytics facing database, as there is no analytical value in user passwords.

> **Note**   As data professionals, we are the last line of defense in protecting this data. It is our responsibility to safeguard personal information, some of which, in the wrong hands, could lead to harm for the individual or individuals involved. Additionally, these individuals entrust our organizations with their data, and may not be aware of the extent of the data held.

Non-PII Personal Data

Personal data is an umbrella term for data pertaining to an individual. In this section, we're going to cover non-PII personal data; information that may not be able to pinpoint a specific individual, while at the same time is specific information about an individual. For the purposes of this chapter, we're going to use the term personal data to refer to this type of data, though it is important to note that in the European Union, the term personal data is used differently and treated very strictly, which we will cover in the next chapter. Non-PII personal data might include more generic information such as

- City of residence
- State of residence
- Gender identity
- Sexual orientation
- Workplace
- Age

Alone, none of these pieces of information would be able to identify one person; however, combined with other information, they could be used to identify someone. Working with non-PII personal data is much more nuanced than working with PII. With PII, it is really cut-and-dried; no one should have access to this data unless it is absolutely necessary to do their job. Non-PII personal data, on the other hand, requires a more subjective approach. When we consider the advances in analytics and machine learning, we must take into account how easily we could stitch data together to identify an individual. As a result, we should treat this data with care by limiting the audience.

Anonymized Data

If you ask someone in the privacy world to define **anonymized data**, they will tell you that no such thing exists. If you were to keep pressuring them to define it, they might eventually break down and tell you that anonymized data is not a simple Boolean yes or no, it is a spectrum ranging from somewhat-anonymous to almost-anonymous. Remember, in math class, when we learned about asymptotes? Think of anonymity as an asymptote. We can keep getting closer and closer, but we cannot achieve full anonymity.

When we work with anonymized data, we need to think, how many of these fields would I need to put together to identify a coworker, a friend, or yourself. The actual answer may realistically be something like three or four fewer fields than you would think. Since we can anonymize data using multiple methods, we will explore the different methods and how the method impacts the way we protect the data.

Tokenized Data

In **tokenized data**, certain personal data fields are obscured such that no single record could be used to identify an individual. This is essentially adding noise to your signal, making it harder to decipher. Data can be tokenized using a few methods, each having their own pros and cons. Regardless of the method, tokenizing data is a secure way to obscure data in a targeted way, allowing an analyst to have access to the data they need without worrying about exposing sensitive information.

Tokenizing with Mock Data

Tokenizing with **mock data**, sometimes referred to as synthetic data, has become increasingly popular in recent years due to advancements in technology. It used to be much more difficult to efficiently generate a bunch of random information that looks the same as data you would expect. The benefit of using mock data is that it feels real. It is easy to see which fields contain addresses, phone numbers, or birth dates. This means that a field with addresses might have lines like

- 1234 Main Street

- 74 Meacobrrdae St.

- 39481 Grant Ave.

Engineers can decide whether they want mock data to be fictional looking as in the Main Street example, or the garbled Meacobrrdae St., or if they want the data to feel more realistic with random numbers followed by a randomly generated street name. The purpose of this data informs this decision. If the data has an internal audience that is aware that the data has been tokenized, then it doesn't particularly matter whether or not the data looks real. However, if the audience is unaware that the data is tokenized, they may believe that a particular user lives at 39481 Grant Ave., and may try to send correspondence accordingly.

Using a One-Way Hash

Tokenizing data using a **one-way hash** is the most common way to tokenize data. This means that once the original data has been obscured by the hash, it cannot be returned to its original state. There are many different hash functions, and the functions themselves come and go. In Figure 2-2, we can see how the original data in combination with a key is used as input to a hash function, which then returns the tokenized data. The tokenized data cannot be returned to the original data since there is not an inverse function for the hash function.

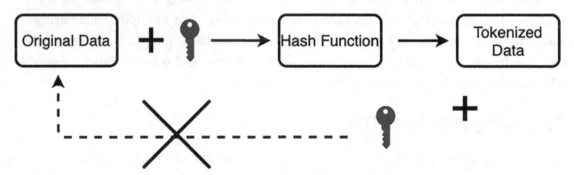

Figure 2-2. *Data that has been tokenized cannot be restored to its original state, even with a key*

Hash functions are used to replace the original data with the hashed value. Every hash function must take one value x and output a value $f(x)$ such that $f(x) = f(y)$ if $x = y$. This means that given the same input, the hash function always returns the same value, or, in other words, the hash function is **deterministic**. A common misconception is that hashing is random. This is not accurate because, by definition, randomness is **non-deterministic**. If a random function were called with the same input ten times, it is likely that there would be up to ten unique values returned by that function.

While hash functions must map exactly one output for a given input, different inputs can share the same output. Hash functions can be one to one, meaning that for every unique input there is a unique output, or they can have a many to one relationship where multiple different inputs map to the same output. When multiple different inputs map to the same output, we call that a collision. Most hash functions have a many to one mapping, and occasionally have collisions. In Figure 2-3, we can visualize the relationship between input and output in a one to one function vs. in a many to one function.

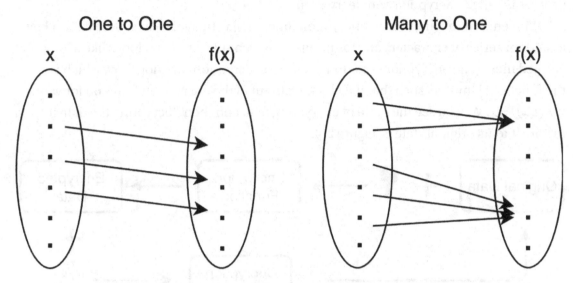

Figure 2-3. *One to one functions map each x to one unique f(x), many to one functions map multiple x values to a given f(x)*

If you are unsure whether or not your hash function is one to one or many to one, it is good to assume that it is one to one when protecting data, and to assume that it is many to one when analyzing data. This means that in the worst case, assume that the hash function is not providing adequate protection for privacy, while at the same time noting that analysis of two different rows with the same hashed value may not represent the same underlying object.

Two commonly used hashes, md5 and sha256, use a **salt**, a series of characters and/or numbers used in some hash functions, in combination with the original data, to come up with a series of characters that represent that source data. Once data is hashed using one of these algorithms, there is no way to get back to the original data. Even checking whether the result of two different hashed values are equal is not a guarantee that the source data is identical as neither function guarantees uniqueness. Sha256 is now considered to be much safer to use than md5, and is the recommended algorithm to use.

Encrypting Data

Encryption is sometimes used synonymously with tokenization; however, there is a very important distinction between them. With tokenization, the mashing-up is unidirectional. That is, once the data has been obscured, it cannot be restored to its original form. With encryption, data can go from something human readable, to a hash of seemingly random characters, and back to its original human readable form. This process is called encryption and decryption.

With encryption, there is an encryption function $f(x)$ that takes in source data, a **key**, a random series of characters and/or numbers, which spits out what looks like a long, garbled mess. The encryption function also has a decryption function $d(x)$, which is the functional inverse, such that if $f(x)$ is the output of the encryption function for value x, $d(f(x)) = x$. We can see the cycle of encrypting data and then decrypting the data to restore it to its original state in Figure 2-4.

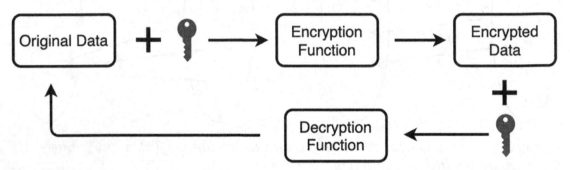

Figure 2-4. *Encryption involves an encryption function, and a decryption function that restores the original data*

This function takes the garbled mess, the key, and returns the original data. The presence of a decryption function makes this data only as safe as the key that protects it. Anyone who has access to that key has the full power to decrypt the data and find themselves in possession of raw, sensitive data. This is why it is recommended to use a one-way hash if the data does not need to be restored to its original value.

We will spend more time on tokenized data later in this book as it can be used in addition to access controls to obscure extra-sensitive fields that aren't necessary for anyone to view.

Aggregated Data

Just as how we can metaphorically break the links between the data and the user to create anonymization by tokenizing data, we can break the links by aggregating the data. Instead of user-level data, we can instead aggregate so the granularity is a population of users. This provides protection against determining an individual from the data. When we use **aggregation**, we need to ask ourselves a couple of questions. We'll explore these through the following example.

Let's imagine we are working with data collected from an employee happiness survey. This survey includes answers from every employee at a company, and the questions target culture-related topics like how included an individual feels, whether or not they feel supported at work, and whether they have the tools to do their job. Each employee answers questions about their manager, their department, and the company as a whole. As a result, we want to be able to aggregate at each of those levels so we can identify areas where these levels can improve.

Can we identify a user from the intersection of multiple groups?

We can imagine that if we compared the results of this survey to results of another survey of employee resource groups, we may be able to identify an individual who has similar answers and is present in both groups. In this example, let's say that there is a team with only one woman on the team, who answered similarly on the employee resource group survey to how she answered on the employee happiness survey. It may be easy to pick out her input from others by seeing that intersection.

What is the smallest group we want to allow in aggregate numbers?

It is very important that when we aggregate data, we think about the threshold for minimum number of members of a group. If we have a newly created team with one team member and one manager, and we aggregate by team, the aggregated data will only include one person's responses. Therefore, in that example, aggregation provided no protection to that individual. Another example is postal codes or postcodes. In some countries like the Netherlands, a postcode could have as few as 20 households, yet we may see 20,000 households in a postcode in Germany. As the aggregate population grows, the individuals are increasingly protected, and when the aggregate population shrinks, individuals may entirely lose their anonymity. Whenever we aggregate, we must set a minimum threshold to prevent de-anonymization.

Internal Non-sensitive Information

Now that we've flushed out some of the most sensitive data you can expect to come across, we need to cover some of the less sensitive data that may not need to be protected much, if at all, within an organization. Non-sensitive internal data can be any information that doesn't involve any of the preceding data types, but that is still restricted to employees within an organization, or those who have signed non-disclosure agreements. Some examples are

- Mappings between group IDs and group names

- The company's organizational structure

- Engineering performance data

Since this data does not include sensitive information about users or employees and has an internal audience, we can consider this to be non-sensitive in most organizations.

Publicly Available Information

Publicly available information is the least sensitive of all data types. Since the data is readily available to everyone, there is no risk of infringing on an NDA or failing to adhere to regulations. As long as the information is readily available for anyone to see, it is considered public. Caution should still be used when this involves individuals. Some examples of publicly available information include

- Map datasets

- Government data

- Research data

- Company names and addresses

- Demographics of municipalities

For the most part, these datasets are public and published with the intent of availability for general use.

Other Sensitive Data

There are a couple types of data that do not neatly fit into our spectrum for one reason or another. Free text fields are a variable type of data – that is, they may contain anything. Financial reporting data is a special case as well because while revenue, profit, and earnings aren't necessarily sensitive data like social security numbers and emails, they may be restricted due to financial reporting regulations.

Free Text Fields

Free text fields can contain any of the sensitive information listed earlier. We need to be cognizant that some text fields may be used in unintended ways. One real world example I've seen is in customer service tickets. In this situation, many customers would put credit card numbers, phone numbers, addresses, etc., in the customer chat we provided on the website. Even though we had separate fields in those tickets for sensitive information like I mentioned, we still found that users would type this in communications. Fortunately, we anticipated this and treated that free form text as having credit card information in it. This meant that we added all comments, emails, and transcriptions to our list of restricted customer data. Potentially sensitive free text fields might involve

- Customer support tickets

- Chat bubble conversations

- Customer surveys

- Reviews

- Customer suggestions

In general, unless you're very sure of the contents of a field, always assume that when things can go wrong, they will go wrong. In other words, assume that the data you handle is more sensitive than the field name suggests.

Financial Reporting Data

Financial reporting data about an organization is a very different type of sensitive data than everything we've covered so far in this chapter. Unlike the previously mentioned types of data, with financial data, we are primarily concerned about users' ability to modify the data rather than view the data. There are still reasons why we might care whether or not individuals can see certain types of financial data at certain times, for example, employees of a public company should not be able to see non-public and material financial reporting data during a trading window. However, we are mostly concerned with an individual's ability to mutate the data to make the financial situation within the organization appear to be different than the reality. Some examples of financial reporting data include

- Revenue for the current quarter

- Advertising spend

- Monthly active users

- Figures related to growth and revenue

Financial reporting data includes standard financial data like revenue and advertiser spend, but also includes metrics that are reported to investors due to being correlated to the aforementioned financial data. In the next chapter, we will cover the regulations involved with financial data and a bit into the history leading up to the regulatory changes.

Key Takeaways

- Sensitive data can describe consumers as well as employees and contractors.

- PII is a generic term for information that can pinpoint an individual.

- Not all PII is created equal; some PII is more sensitive than others.

- Non-PII personal data can be stitched together to identify an individual.

- Anonymized data is not necessarily anonymous and, therefore, safe.

- Data can be tokenized in different ways.

- Tokenization is one way and cannot be reversed.

- Free text fields can contain sensitive information.

- Err on the side of caution with data types.

In the next chapter, we will dive deeper into the regulatory requirements around these different types of data.

CHAPTER 3

Data Privacy Laws and Regulatory Drivers

Privacy is regarded as a fundamental human right; however, that wasn't always the case. In the United States, privacy was not a constitutionally protected right – there is no privacy clause in the Constitution. The history behind privacy becoming a fundamental right in the United States started with the feminist movement in the 1960s. Women wanted the right to control their reproductive destiny within their marriages and states, including Connecticut banning contraceptives. This culminated with the Supreme Court case *Griswold v. Connecticut (1965)*. The Supreme Court ruled in a 7–2 decision to overturn Connecticut's law, that there exists a "right to marital privacy" stating that the Due Process Clause in the fourteenth amendment implies that privacy is a right guaranteed to all citizens in the United States by the federal government as well as state governments. This Supreme Court decision not only paved the road for women in higher learning and the workplace, benefiting women like myself, but also serves as the basis for privacy as we know it in the United States.

Internet Privacy

Big data moves at a blistering speed. Anyone working in the industry knows that best practices are constantly shifting, and new technologies keep advancing. In contrast, legislative bodies move slowly and deliberately, while at the same time, many of these legislators do not have high technical proficiency. The Internet is filled with hilarious videos of gaffes our elected officials have made while trying to understand the Internet. The late US Senator Ted Stevens famously referred to the Internet as a "series of tubes," explaining that an email he was sent was stuck in a clogged and full Internet pipe.

© Jessica Megan Larson 2022
J. M. Larson, *Snowflake Access Control*, https://doi.org/10.1007/978-1-4842-8038-6_3

During United States congressional hearings with Google and Facebook executives, many legislators showed that they fundamentally did not understand the technology. As a result, we are likely to see less comprehensive regulations with ill-defined policies, and a whack-a-mole style approach to amending these original laws.

As data collection becomes a standard practice for any organization with an online presence, consumers are becoming increasingly concerned about the safety and use of their data. In the United States, organizations like the IEEE (Institute of Electrical and Electronics Engineers), the Electronic Frontier Foundation (EFF), and the ACLU (American Civil Liberties Union), in partnership with many other nonprofit groups, fight tirelessly for consumer protections.

In this chapter, we are going to cover a selection of commonly applicable data privacy laws and regulations. The intent of this chapter is to familiarize yourself with some of the different regulations that are out there. We can see a high-level overview of the laws we will cover in this chapter in Figure 3-1.

Regulation	Where is this enforced?	What types of data are restricted?	How is data restricted?	What is the Intent?
GDPR	European Union	Personal Data	Collection Access Deletion	Protect Consumer Privacy
APPI	Japan	Personal Data	Collection Access Deletion	Protect Consumer Privacy
CCPA	California, USA	Personal Data	Collection Access Deletion	Protect Consumer Privacy
SOX	USA	Financial Data	Modification	Prevent Financial and Securities Fraud
HIPAA	USA	Personal Health Data	Access	Protect Personal Health Privacy

(Left margin: "More Strict" with upward arrow transitioning to "Less Strict" with downward arrow)

Figure 3-1. *Comparing data regulations from this chapter*

Since these are complex regulations that require a holistic compliance approach, we're going to focus on the access control in data warehousing aspects.

Disclaimer   I am not an attorney, and the following analysis does not constitute legal advice. I am presenting my analysis of these regulations based on my experience as a data engineer. For legal advice, please seek legal counsel in your jurisdiction.

GDPR

GDPR, or rather, the General Data Protection Regulation is the most well-known set of data privacy laws, covering all individuals who are residents of the European Union (EU) member states. It carves out many different protections to consumers that did not exist previously and applies to all companies that do business in the EU. Certain countries like Germany and France have laws that strengthen the protections granted under GDPR to include additional types of data. This set of laws fundamentally changed how technology companies do business in Europe and was an awakening for the industry. Many large companies now have entire teams of lawyers and developers on staff dedicated to maintaining GDPR compliance. GDPR posits that consumers have a fundamental data privacy right that cannot be taken away, and that the intention of GDPR was to protect and extend those rights to the digital world. The second main overarching concept is that of data protection by design and by default.

Definitions

Before we dive into the specific consumer protections, we need to establish some definitions. The following terms have more broad connotations in typical use than they do under GDPR. These terms are used frequently throughout the text of GDPR, and will be referred to in the following section.

- **Data subject** – A data subject is a person who is the subject of personal data held by an organization.

- **Pseudonymization** – A type of anonymized data that cannot be traced back to the data subject without the use of additional data, where the additional data is separated and held in such a way that an individual cannot be identified.

Who Is Affected?

GDPR is intended to protect all residents located within the EU. The organizations that must comply with GDPR are less cut-and-dried, however. In general, any organization that processes data of residents of the EU is subject to the regulations imposed in GDPR. Not-for-profit organizations and those who do not engage in economic activity are exempt, as well as organizations with fewer than 250 employees. Companies slightly larger may not be required to adhere to all of the requirements GDPR prescribes. Additionally, companies located in the EU that process data of individuals outside of the EU are still bound by these laws.

Special Category of Personal Data

GDPR outlines a **special category** of particularly sensitive and restricted personal data, which they refer to as a "special category" of data. As shown in Figure 3-2, this data includes information about race, ethnic origin, political opinions, religious or philosophical beliefs, trade union membership, genetic data, biometric data for the purpose of identifying an individual, health data, and data concerning an individual's sex life or sexual orientation.

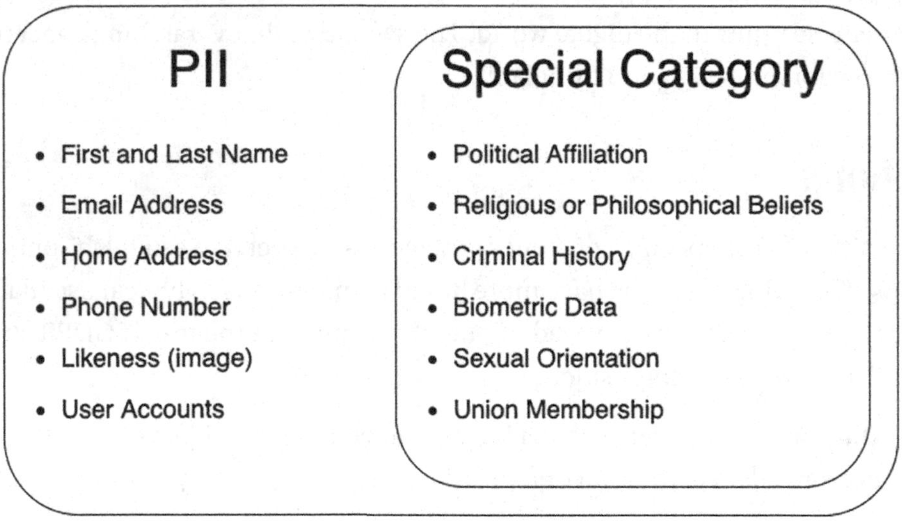

PII	Special Category
• First and Last Name	• Political Affiliation
• Email Address	• Religious or Philosophical Beliefs
• Home Address	• Criminal History
• Phone Number	• Biometric Data
• Likeness (image)	• Sexual Orientation
• User Accounts	• Union Membership

Figure 3-2. *GDPR's Special Category of PII compared to standard PII*

This special category of data is considered the most sensitive through the GDPR lens. This type of data is restricted such that processing of this data is prohibited, with a few exceptions. One exception is if the data subject explicitly consents to processing of this data. Another exception is if the organization that holds this data must process this data for critical business reasons that benefit the data subject, such as an organization operating in the medical industry, or a credit issuing company. A not-for-profit organization that serves a similar cause may process the data, such as a not-for-profit organization serving the LGBTQ+ community processing data related to the sexual orientation and identity of its members. If an individual is involved in legal matters, that carves out an exception as well. Additionally, research in the public's interest that involves these matters is also exempt.

Data Processing Principles

GDPR includes provisions around how data can be collected, processed, stored, accessed, and updated. The provisions we are going to focus on in this chapter are data minimization, storage limitation, and integrity and confidentiality. These categories directly impact access control.

Data Minimization

Organizations are asked to minimize the amount of data they collect pertaining to data subjects. If a personal data field doesn't have analytical value, then, under GDPR, it doesn't belong in an analytics database. Data collected about data subjects must be "adequate, relevant and limited to what is necessary in relation to the purposes for which they are processed."

Storage Limitation

PII should only be held in a format allowing the identification of an individual for the minimum amount of time required for the purpose of processing the data. Once that purpose has been fulfilled, the ability to identify an individual must be stripped. This is where pseudonymization comes in.

Integrity and Confidentiality

Personal data held about data subjects must be stored in such a way as to prevent unauthorized access. Employees should only be able to access the data they need for the intended processing purpose and should not be able to modify or delete the data without authorization. Restricting access also helps prevent against data breaches.

Handling Data with GDPR Compliance

Since the majority of GDPR's requirements fall outside the scope of an analytical database, handling data with GDPR compliance is within reach. We must keep all PII separate from the rest of the data, and it must only be accessible to those who need access. When the identification of an individual is not critical, PII must be pseudonymized such that it is not possible to deanonymize. Finally, if personal data does not have a purpose for processing, then it should not exist in its raw, unaltered form in an analytics database. Whenever possible, PII should be tokenized, encrypted, and/or aggregated.

APPI

Established in 2003, nearly a decade before the spread of big data, **APPI**, or the Act on the Protection of Personal Information, was drafted as Japan's first consumer privacy act. While APPI was the first of the consumer data privacy acts to go into effect in 2005, it is not nearly as well known or as frequently referenced in the industry as GDPR. This is due in part to the fact that Japan does not have as much power on its own as a single country with the 11th largest population compared to the combined 27 member states in the EU.

Since APPI was initially drafted before the widespread use of social media and before the spread of targeted and personalized advertising, most of the original regulations are concerned with the consent of the individual and the disclosures made by the organization, as opposed to how the data is stored and processed. This made it less applicable to data professionals than GDPR and others since the onus of compliance largely fell on attorneys and terms of service agreements. However, in response to GDPR, APPI was amended to include many of the same protections. This was done in combination with the founding of the EU-Japan Economic Partnership Agreement. EU residents now have the guarantee that when their personal data is transferred to Japan, they will receive all of the same protections.

Definitions

Like GDPR, APPI defines certain terms a little differently than they may be used in other places. The following term is defined and used in APPI to describe a category of data:

- **Personal information** – PII specifically pertaining to living individuals in Japan

Who Is Affected?

The scope for APPI is similar to the scope of GDPR. Any entity that handles personal information of living individuals in Japan must adhere to APPI. Governments and government agencies are exempt from these regulations.

Handling Data with APPI Compliance

Similarly to GDPR, the majority of APPI falls outside the scope of an analytics database. In the case of APPI, most of the regulations apply to the acquisition of the information. Organizations must specify a purpose under which the data is processed and must disclose that purpose to the individuals whose information they're collecting. APPI also requires that there are necessary and proper protections against data loss. We can infer that we can accomplish this by creating silos for user data that is limited to those who absolutely need access. The last major piece that APPI requires is adequate supervision over those who have access to personal information. To adhere to this, in addition to restricting data access, we can also create guard rails preventing certain actions, and we can log all queries against this data. Since the revised APPI includes all the same protections as GDPR, we can further include the policies we outlined in the GDPR section.

CCPA

The California Consumer Privacy Act, or **CCPA**, is an act modeled after GDPR but for a smaller audience – roughly 10% of the population – the California residents. The Golden State has earned a reputation for leading the pack with some of the strongest consumer protections across the United States. CCPA differs from GDPR in that GDPR requires data privacy by design and by default, whereas CCPA requires consumers to opt-out from their data being sold to a third party. Essentially, we can think of CCPA as *Diet GDPR*.

CCPA focuses heavily on personal data that is sold to third parties and requires that organizations provide a way for consumers to opt out of having their data sold. Additionally, this act allows users to opt out of certain cookies and tracking, requiring organizations to break down the different purposes for different cookies. This act also prevents discrimination against consumers who choose to opt out of having their information sold. If a consumer opts-out, they must be able to have the same experience on the platform as they would otherwise, with some exceptions. California residents can also enjoy their right to deletion, and the right to access the information collected about themselves.

Since the passage of CCPA, California passed the California Privacy Rights Act of 2020 (**CPRA**), strengthening consumer protections and aligning California more closely to the EU. CPRA nearly reaches parity with GDPR by adding additional rights for consumers, including the right to correct outdated or inaccurate information, while at the same time increasing transparency requirements. It also steeply increases penalties against organizations who violate the data privacy rights of minors. This update also includes defining a new and more sensitive category of personal data like we see in GDPR. In Figure 3-3, we can see the different types of personal data outlined in CCPA, including newly defined categories, probabilistic identifiers, and sensitive personal data.

Unique Identifiers	Sensitive Personal Data	Probabilistic Identifiers
• First and Last Name	• Govt Identification Numbers (i.e., SSN)	• Unique Browser Identification
• Email Address	• Passwords	• Shopping Behavior
• Home Address	• Precise Geolocation	• Neighborhood
• Phone Number	• Biometric Data	• Age
• Likeness (image)	• Sexual Orientation	
• User Accounts		

Figure 3-3. *CCPA outlines three types of personal information*

Definitions

Though CCPA makes use of frequently used terms to describe types of data through the lens of data privacy, the law additionally defines a couple unique terms. The following terms are defined in CCPA and the revision, CPRA:

- **Probabilistic identifier** – Data that can be linked to an individual through probabilistic inference with more than 50% certainty

- **Unique identifier/Unique personal identifier** – PII that also includes probabilistic identifiers

Who Is Affected?

Any organization that has consumers, employees, or service providers in California or operates within California is affected.

Handling Data with CCPA Compliance

CCPA mandates that personal information be protected against unauthorized access. We can think of authorized access as someone accessing information they need to do their job. Unauthorized access is anything outside of that. This means that like with GDPR, we need to set up strict boundaries around these unique personal identifiers. In some cases, separating PII from non-PII is sufficient, but in some more complex use cases or in larger organizations, higher granularity is necessary. It is worth noting that CCPA includes probabilistic identifiers in its definition of PII, requiring that information that likely points to an individual be treated with the same respect as uniquely identifying information. This can mean information like a device or a browser, where it is more likely than not that the data identifies an individual or a family, but can also be expanded to include less clear links like shopping behavior. Since the definition of probabilistic identifier includes any adjacent data, this could include a pretty broad category of data.

Additionally, with the passage of CPRA, which goes into effect in 2023, we need to take extra care when dealing with the new category of sensitive personal information. In most cases, this data should not be placed into an analytics database in its raw unaltered form, and if it is necessary to include it, it should be tokenized or encrypted.

US State General Privacy Regulations

While there was momentum in the consumer privacy world, the United States Federal government faced unprecedented and distracting challenges. As a result, the federal government hasn't had much to say in response to GDPR. In this vacuum, and following California's lead, Virginia passed the Consumer Data Protection Act (CDPA). This law was modeled after GDPR and CCPA, taking most of the provisions the two have in common. Instead of pseudonymization or anonymization, CDPA uses the term de-identification, and it outlines much of the same handling of this anonymized data. Less than a month later, Colorado passed a similar law, the Colorado Privacy Act (CPA). This law follows suit and grants many of the same rights to residents of Colorado.

The wave of data privacy laws also includes many states in the process of passing legislation; New York, Maryland, Massachusetts, Hawaii, and North Dakota have introduced data privacy bills. These bills largely resemble CCPA and aim to mirror the same protections. Some of these have been held up and have already failed or are unlikely to pass; however, the trend suggests that this will continue, and more states will follow suit.

SOX

The Sarbanes-Oxley Act, or **SOX**, is a federal act in the United States designed to ensure that financial reports from public companies are accurate and specifically that these financial figures are protected against tampering and other fraudulent behavior. In the wake of the Enron crisis, lawmakers in the United States designed and passed this act of legislation that added criminal charges for financial fraud in large corporations. Additionally, this legislation outlined steps that organizations must take to prevent fraud, specifically that internal controls must be placed.

Definitions

SOX deviates strongly from the regulations we've talked about so far in this chapter, and therefore has entirely different definitions. Here, our definitions refer to our business practices as well as our data infrastructure. The actual components affected by both of these terms will vary from organization to organization.

- **Internal controls** – A set of rules, processes, and practices around how data is handled, transformed, and accessed put in place to prevent and identify fraudulent behavior

- **SOX scope** – All systems involved in moving, storing, or manipulating SOX-controlled data

Who Is Affected?

Corporations based in the United States that are considered issuers by the Securities and Exchange Commission (SEC) must adhere to the regulations in SOX. Essentially this means corporations that are publicly traded in the United States, or that are publicly traded in another country but operate within the United States. If your organization is mature enough for this to apply, your organization will most likely have a legal team on staff that will let you know.

Handling Data with SOX Compliance

To adhere to SOX regulations, we must have robust internal controls that allow us to track all changes made to any system that is in SOX scope. This basically means that we must have processes for dealing with financial reporting data from the source (or more likely sources), all the way to the final report that's made public and sent off to regulatory agencies and investors.

SOX Scope

When we speak about SOX scope, we are referring to all of the internal and external tools involved in handling SOX data. These systems must have strict access control, not so much for individuals reading the data, but protection against individuals modifying the data. Since SOX is primarily concerned with preventing fraud, an individual's ability to read data isn't as much of an issue. As shown in Figure 3-4, an example pipeline might use a cloud-based server to grab data from a point-of-sale system, dump the data into a cloud-based flat file storage bucket, copy that data into a Snowflake database, and finally import the data into an accounting software product for the final report.

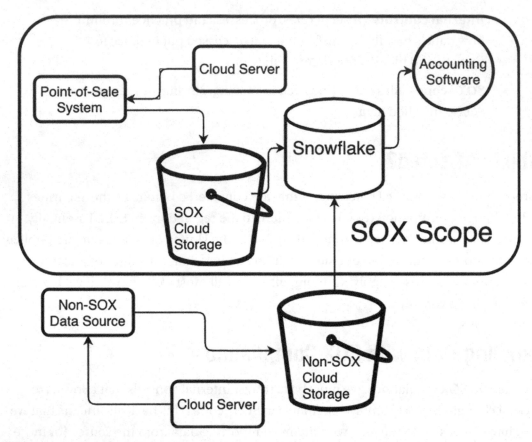

Figure 3-4. *An example of SOX infrastructure and scope vs. out of SOX scope*

All of those components involved in the SOX pipeline are in SOX scope and are therefore bound by internal controls. The elements of infrastructure not involved in the process that still interact with these systems are out of scope.

Creating Financial Reports

To start to think about the process of adhering to SOX, we must understand fundamentally what is happening when we create a financial report. If we were a retail company, we wouldn't just take the raw data of all transactions for the year and send it off. No one would be able to understand that data, it wouldn't tell any story, and it would be much too large for someone to pick through. At the heart of creating a financial report is manipulating the data. While we don't want to provide a list of millions of transactions, the sum or total dollar amount resulting from those sales would be an important figure to report. To do this, we aggregate the transactions and report out the sum. However, there may be situations in which we must first clean the data before aggregating, or we may find that to report certain figures, the calculation is much more complicated. The calculation might even change due to changing business logic or changes in accounting regulations and tax law.

When we create these rules and definitions for figures, we must track and document exactly how we are getting to these numbers. When we change how we calculate these numbers, we must carefully document why we changed it and how we changed it. We also must have certain individuals approve these changes. Every single one of these steps must be documented and tracked. Remember when we talked about SOX scope? Where we document and approve these changes is also part of SOX scope. This might mean a ticketing system for requesting access, or an internal accounting wiki page including definitions.

Restricting Access

Now that we understand what is in scope and why SOX scope is so expansive, we need to understand the controls that must be in place for all of these systems. In an ideal world, no single individual would have the ability to modify any of the underlying data that powers a financial report. In practice, we do not have the technology to make sure that is possible. We can however, mitigate risk in a few ways. We can utilize an entirely separate set of infrastructure for SOX data, restrict this type of access to as few individuals as possible, and we can track every single change made. To keep the separation clear, we can use a separate staging environment, and a separate account, database, or schema in Snowflake. For the individuals that must have access to the data, they are required to complete SOX training and are liable in the event they violate internal controls, which means that they could be criminally prosecuted. In addition to restricting access to a small group of people, we must have a robust system in place for users to request access to SOX datasets, and for SOX approvers to accept or deny their request. This can be done easily in a ticketing system which can track the rationale behind granting the access and the timeframe a user had access to this data. To track every single change made to the dataset, we must use systems that allow us to track those changes. In the case of Snowflake, the typical course of action is to log all queries against these datasets to an external logging system with controls against modification and deletion.

HIPAA

HIPAA, or the Health Insurance Portability and Accountability Act, regulates what certain types of US organizations within the healthcare industry can do with personal health information (PHI) pertaining to individuals. A common misconception about HIPAA is that it applies to anyone handling health data, which is incorrect. A soccer club requiring players to be fully vaccinated to play does not have to follow the guidelines outlined in HIPAA. This regulation is primarily concerned with regulating how entities in healthcare transfer PHI in order to facilitate care for an individual, and the process by which the information is disclosed.

Definitions

HIPAA again deviates from the rest of the regulations covered so far in this chapter, as it is aimed at the different organizations that may be involved when an individual receives healthcare. The following definitions explain which organizations must comply with HIPAA:

- **Covered entities** – Health plans, healthcare clearinghouses, healthcare providers, and healthcare business associates that are required to adhere to HIPAA

- **Health plan** – A plan that provides or pays the cost of medical care, including a group health plan, a health insurance issuer, Part A or Part B of Medicare, and many other similar entities

- **Healthcare clearinghouses** – An entity including a billing service, repricing company, community health management information system, and "value added" networks

- **Healthcare providers** – An entity that provides medical care or treatment

- **Healthcare business associates** – Entities or individuals who work with the preceding entities in a way where exchanging PHI electronically is required

Who Is Affected?

Covered entities are required to adhere to HIPAA. Covered entities include health plans, healthcare clearinghouses, and healthcare providers. In this case, health plans is referring to any entity providing a health plan, which we typically refer to as health insurance companies. Healthcare providers would include any entity providing health treatment to individuals. Healthcare clearinghouses is a large umbrella term that covers any entity that is related to insurance companies or healthcare providers. This could include any entity that interfaces with these companies, such as a company that helps manage relationships with a health insurance company.

Handling Data with HIPAA Compliance

To comply with HIPAA, covered entities must secure access to PHI physically, administratively, and technically. To secure the data physically, passwords must be required for access to workstations, and those workstations must automatically log users out after a period of inactivity. To secure data technically, through software, all PHI must be fully encrypted at rest and in transit. Additionally, only authorized users should be able to decrypt and use or disclose data. Authorized users means users accessing personal health data in order to carry out treatment, payment, or health care operations who are explicitly given consent by the individual. To accomplish this, every employee or provider accessing this data must have their own distinct user account and password. This is crucial because without this, activity cannot be traced back to an individual.

Because of the rigidness of HIPAA's regulations, I would not recommend directly querying PHI in Snowflake, and instead using Snowflake to power a highly secure health care tool. This means encrypting the data before it hits Snowflake and decrypting in the third-party tool.

Future Regulations

When we look to the future, I would predict that we will see a number of laws emerge that resemble GDPR – the gold standard of data privacy. I expect that we will see a federal US equivalent by 2025; though due to lobbying power and pushback from technology companies, it will ultimately provide weaker protections than its EU counterpart. In that time, I expect that the EU will greatly strengthen GDPR leaving the rest of the world to catch up again. Additionally, I would expect that the EU will add more countries to their growing list of countries they recognize as providing adequate protections, typically coupled with free trade agreements between the union and those countries.

Generalizing Data Privacy Regulations

It is possible that none of these regulations or anticipated regulations affect your company; however, it is much more likely that one or all of these affect your company.

I recommend generally adhering to all of these regulations with respect to the analytics database as much as your organization allows because the landscape looks like these policies will only become more rigorous as time goes on, and the effort to put everything in place correctly the first time will be easier than doing half of it now and half of it when the clock starts ticking. This would be in addition to the required compliance steps. Additionally, most of these practices will make your analytics infrastructure more robust.

We can take the following steps to protect personal data at our organization:

1. **Identify and categorize all data** – What fields do we consider personal data, PII, or the special category of PII? What tables include these fields?

2. **Create policies around working with sensitive data** – Should users ever be able to see raw email addresses? When should we tokenize data and when is restricting access enough?

3. **Separate out sensitive data from non-sensitive data** – We can do this when we create our data models, or we can evolve existing ones.

4. **Lock down access to sensitive data** – Maybe we put this data in a separate schema, maybe we put it in a separate table; either way, exclude as many eyes as possible.

5. **Tokenize or remove special category data** – This data has little analytical value and needs to be protected at all costs.

By following these steps in addition to the steps we need to take to achieve compliance, we can ensure that our analytics database is future-proofed against the next wave of regulations.

Key Takeaways

In this chapter, we covered the different laws that affect how we store and protect data in Snowflake. GDPR, APPI, and CCPA aim to protect consumer privacy, and generally affect anyone doing business in those jurisdictions. SOX aims to prevent financial fraud in the fallout of the Enron crisis. HIPAA protects individual's health data from improper handling by the organizations that facilitate healthcare. All of these regulations, at their core, aim to protect individuals, whether they're consumers, patients, or shareholders. Consumers have an appetite for these data privacy laws, and it is likely that there will be an influx of new and revised regulations in the future.

- Most data privacy laws apply to read access.

- The majority of data privacy laws protect residents in a jurisdiction.

- PII can be broken down into a more sensitive special category.

- PII should be used and accessed on an as-needed basis.

- The special category of PII shouldn't be accessible to anyone.

- GDPR is the frontrunner and model data privacy law.

- SOX compliance involves write access and all involved systems.

- HIPAA only applies to covered entities – mostly health care providers.

- There will be future laws, prepare for them.

Sources

Griswold v. Connecticut: www.plannedparenthoodaction.org/issues/birth-control/griswold-v-connecticut

GDPR English Translation: https://eur-lex.europa.eu/legal-content/EN/TXT/?uri=CELEX%3A02016R0679-20160504&qid=1532348683434

APPI English translation: www.cas.go.jp/jp/seisaku/hourei/data/APPI.pdf

Japan – EU data union: https://ec.europa.eu/commission/presscorner/detail/en/IP_18_5433

California CCPA: https://leginfo.legislature.ca.gov/faces/codes_displayText.xhtml?division=3.&part=4.&lawCode=CIV&title=1.81.5

Virginia CDPA: https://lis.virginia.gov/cgi-bin/legp604.exe?212+ful+HB2307ER

Colorado: https://leg.colorado.gov/sites/default/files/documents/2021A/bills/2021a_190_rer.pdf

SOX: www.congress.gov/bill/107th-congress/house-bill/3763/text

HIPAA: www.hhs.gov/sites/default/files/ocr/privacy/hipaa/administrative/combined/hipaa-simplification-201303.pdf

Ted Stevens Quote: www.nytimes.com/2006/07/17/business/media/17stevens.html

CHAPTER 4

Permission Types

Gone are the days of simple access control paradigms, at least for data warehouses. We have increasingly complex datasets and as we saw in our last chapter, we have increasingly complex regulations to meet. As a result, we needed to adopt a more robust system of privileges within data warehousing. In order to manage this effectively, we need to thoroughly understand the privileges that exist, and we need to create groupings for the privileges to simplify granting access so that we can manage them at scale. Since these privileges get very specific, we need to understand them so that we know what we're granting. By creating groupings, we're not only making it easier to understand, we're also eliminating the need to manually identify and grant fifteen different privileges to a role every single time.

Before we dive into the groups, we're going to look at the fundamental building blocks of access control.

Permission Type Concepts

Linux has set the standard for excellence in open source. Virtually everything runs on Linux, and, as a result, it has great influence over everything computing. Permission structures are no exception. For these reasons, I think it is worthwhile to spend a little bit of time reviewing how Linux structures access control. The standard permission paradigm breaks down access into three categories: read, write, execute, which are broken down based on ownership. Read access means a user is able to view the object and its contents. Write access means that a user can modify or delete an object. Execute means that a user can execute a program. Object ownership delineates the read, write, and execute permissions.

When viewing files in a terminal window, we can see the permissions for files and directories. Figure 4-1 shows the structure of a permissions string in Linux.

© Jessica Megan Larson 2022
J. M. Larson, *Snowflake Access Control*, https://doi.org/10.1007/978-1-4842-8038-6_4

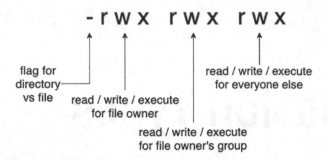

Figure 4-1. *Permissions in Linux*

As shown in Figure 4-1, permissions are done based on scope relative to the owner of the file. The first string of permissions is enforced against the owner of the file. The second section is enforced against the group the owner belongs to, and the third section applies to everyone outside of the group.

When the permission string is filled out, each section will have either a r/w/x or a - for each read, write, execute, respectively. If a file is set to read, write, and execute for the owner, the owner's section would look like rwx. Let's say for this same file, we grant the group read and execute access, but not write access, we would give that section r-x. For everyone else, we grant read only access, r--. Putting this together, we have -rwxr-xr--, which then is turned into 754 using binary for each of these sections. We can think of these different sections like roles in Snowflake, each section represents the privileges granted to a different group of users.

Files on a computer do not live in a vacuum, they live in a hierarchy. Every file exists in a directory, and in order for a file to be accessed, the user who is accessing the file must have read and execute permissions for the directory. This hierarchy can be further expanded to include access to the machine; if a user cannot authenticate and access a machine, they will not be able to access the file or the directory. When we think about the hierarchy of permissions, we must take into account the path we must take to access these files. We traverse the hierarchy from top down, starting from the initial access to the machine, through the directory tree, down to the file itself. If at any point between the highest level of access to a system to our file we don't have access, then our traversal stops there, and we cannot access the file. Permissions after this breaking point are essentially moot, they will not be invoked unless that breaking point is resolved.

These concepts will help guide how we think about permissions in Snowflake. Later in this chapter, we will return to this.

Privilege Scope Categories

In the last section, we introduced the hierarchy of permissions within Linux. Since we've covered a more basic example, let's look closer at the different categories of privilege scope in Snowflake. We're going to move from highest level in the hierarchy to the lowest levels.

Global and Account

Global and Account Privileges are permissions that exist that affect the entirety of an organization or account. These privileges affect objects that exist at the highest level like the ability to create databases, shares, accounts, warehouses, or security integrations, and privileges associated with managed reader accounts. This also affects things we may not consider to be objects like the ability to create network policies, the ability to manage grants, and the ability to monitor usage. All of these permissions have a cascading effect on the rest of the organization in Snowflake.

Databases, Warehouses, and Other Account Objects

Account Object Privileges are permissions that affect objects that live in the account bucket, rather than objects that live in a schema within an account. Account objects include databases, warehouses, users, resource monitors, and integrations. Objects like these are separate from other types of objects in an account because they don't inherit permissions from a higher level in the account; if a user is granted the ability to modify a database, then as long as they have a Snowflake user account within this Snowflake account, they will be able to modify the database. Grants at this level will not necessarily affect all downstream objects, that is, if a user's role is granted usage on a database, that same role must also be granted usage on the underlying schemas in order to see those schemas appear.

Schemas

Schema Privileges are the next highest level of privileges in the Snowflake database. Schemas are the smallest bucket we have to store objects. If we think of our Snowflake database as a Linux file system, databases are folders containing only nested folders, and schemas are folders that contain files with data. As a result of schemas holding many data objects, we need to be very careful about how we allocate access within schemas. The main type of permission we have at the schema level is the ability to create objects, including, but not limited to, a table, view, file format, stage, or pipe. In addition to this, we have the permissions to see that a schema exists, modify a schema, or own a schema.

Schema Objects

Schema Object Privileges include all permissions around objects that live inside schemas. While we use schema permissions to allow a user to create objects, all other object permissions must be done at the object level. This means that we can create a schema level permission for creating tables, but we must grant the ability to read data from or modify a table separately through object permissions. Later in this chapter, we will cover bulk grants on schema objects using the all modifier and the future grants concept. While there is some variation in the different privileges that exist for each type of object, we always see the four CRUD privileges, create, read, update, and delete. In Snowflake, they map to the following privileges, create maps to CREATE, read maps to SELECT, update maps to UPDATE or INSERT in some cases, and delete is granted through OWNERSHIP. In the next section, we will dive into what these different types of privileges mean.

Snowflake Permission Types

Snowflake provides very granular privilege types; each type of object has its own set of privileges, which allow an administrator to create very targeted roles. However, with this freedom comes the complexity which can be a bit difficult to understand, and even more difficult to manage. In this section we will cover the main privilege concepts in Snowflake.

General

Many of the different privilege types apply to more than one type of object in Snowflake. The privileges that apply to all or many types of objects include the following:

- OWNERSHIP – Only one role can own an object in Snowflake. The ownership privilege allows a role to rename, delete, modify, and allocate access to a resource. Ownership defaults to the role that created the object, but ownership can be granted to another role, effectively transferring ownership.

- ALL – The "all" privilege grants all applicable privileges for an object to a role, with the exception of ownership. Certain objects may also have a privilege that is excluded from the all privilege as well.

- CREATE <OBJECT TYPE> – The create privilege must be granted one level above the object to be created.

- MODIFY – The modify privilege allows a role to modify the parameters specified on an object.

- MONITOR – The monitor privilege allows a role to view the details of an object. This can be done using the DESCRIBE keyword.

- USAGE – The usage privilege allows a role to utilize a resource using the USE keyword, and to list objects within a resource using the SHOW keyword.

When granting CREATE for a certain object type, the grant must occur in the object above the object to be created. For example, as we can see in Figure 4-2, to grant the ability to create a view, it must be granted on a schema, since that is where the view lives.

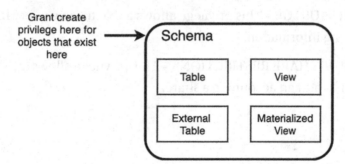

Figure 4-2. *Create privileges must be granted on the container of the object*

When we visualize this, it makes sense that CREATE VIEWS should be granted at the schema level, as that is where views exist.

Global and Account

Global privileges include a few specific abilities that affect the entire account, in addition to the ability to create global level objects. The CREATE privilege can be used to create the following objects in the global space: **account, database, integration, tag, network policy, role, share, user, warehouse, and data exchange**. Global privileges also include the following:

- APPLY <OBJECT TYPE> – This privilege allows a role to apply a column masking policy or a row access security policy on a table or view, or to apply or remove a tag on an object.

- ATTACH POLICY – This privilege allows a role to use a network policy on an account.

- EXECUTE TASK – This privilege allows a role to trigger a task owned by the same role.

- IMPORT SHARES – This privilege grants the ability to see shares and to create databases from them.

- MANAGE GRANTS – This privilege allows a role to grant or revoke privileges on any object regardless of privileges or ownership on that object.

- MONITOR EXECUTION – This privilege allows a role to view the status of all pipes and tasks in the account.

- MONITOR USAGE – This privilege allows a role to view query history and billing information.

- OVERRIDE SHARE RESTRICTIONS – This privilege allows an account to add an account to a share.

Databases

Since databases are used to store schemas, which then in return store data, we can imagine that database level permissions may mainly affect schemas. Since schemas are the only type of object that directly lives in a database, the only object that can be created at the database level is a schema, and thus the only CREATE privilege on a database. The following privilege can be granted at the database level for databases created from a secure data share:

- IMPORTED PRIVILEGES – The ability to access a shared database

Schemas

Schemas are where everything starts to get interesting since data is organized into objects within a schema. The following objects can be created at the schema level:

- Table (external or otherwise)

- View (materialized, secure, or otherwise)

- Masking policy

- Row access policy

- Stage

- File format

- Sequence

- Function

- Pipe

- Stream

- Task

- Procedure

As mentioned in the "General" section, USAGE is the privilege that allows a role to view the objects within an object, if roles do not have the usage privilege on a schema, they cannot access tables or views within that schema, and they cannot see whether or not they exist.

Tables and Views

Tables and views include some of the same privilege types. There are no objects within a table or view that can be created using the CREATE keyword. Let's dive into the different types of table and view privileges:

- SELECT – The ability to directly query a table or view; this allows them to literally invoke the SELECT keyword.

- INSERT – The ability to insert new rows of data into a table using the INSERT keyword, in addition to the ability to run an ALTER TABLE command.

- UPDATE – The ability to change the value in one or more fields in one or more rows in a table using the UPDATE keyword.

- DELETE – The ability to remove rows from a table using the DELETE keyword.

- TRUNCATE – The ability to drop all rows from a table using the TRUNCATE keyword.

- REFERENCES – The ability to create another object that uses this table or view as a foreign key constraint using the REFERENCES keyword; this also includes the ability to view the table and its structure.

Warehouses

Warehouses, like tables and views, are a fundamental object type that cannot be a container for any other type of object. As a result, there are no create privileges at the warehouse level. While the following privileges have been described in the "General" section, the way these privileges affect a warehouse differ from the definition for other object types. These privileges on a warehouse are described here:

- MONITOR – The ability to see the usage and queries run on a warehouse

- OPERATE – The ability to start, stop, resume, or suspend a warehouse

- USAGE – The ability to run queries using a warehouse's compute

Granting and Revoking Privileges

Earlier in this chapter, we covered the different privileges to grant to certain types of roles. The next piece of the puzzle here would be the actual granting and revoking of these privileges. While this seems straightforward, there are certainly pitfalls. A best practice to keep in mind is logging all grants and revokes in a version control solution like GitHub.

Granting Privileges

When we give someone the ability to do something, we say that we are granting a privilege. Granting privileges to roles is the easy part. We can grant privileges on objects like

```
GRANT <PRIVILEGE> ON <OBJECT TYPE> <OBJECT NAME> TO ROLE <ROLE NAME>;
```

Or for privileges that aren't specific to a particular object, we can issues a grant like

```
GRANT <PRIVILEGE> TO ROLE <ROLE NAME>;
```

In practice, we might grant account or global privileges to a role like

```
GRANT MANAGE GRANTS TO ROLE SNOWFLAKE_DBA;
```

We can also include arguments and modifiers, which we will dive into in the next section. It is important to note that we typically need to grant more than one privilege at a time to give a user access to a particular object. For example, if we have a table called EMPLOYEES in our PROD database and in our HR schema, we might grant a user with the role SNOWFLAKE_USER the ability to read data from that table like

```
GRANT SELECT ON TABLE PROD.HR.EMPLOYEES TO ROLE SNOWFLAKE_USER;
```

However, if this is the first privilege allocated to this role, the user will not be able to view data in that table. We must additionally grant access to the containers above that table as well.

```
GRANT USAGE ON DATABASE PROD TO ROLE SNOWFLAKE_USER;
GRANT USAGE ON SCHEMA HR TO ROLE SNOWFLAKE_USER;
```

Great, now the user has access to the table, can see the fields, and start to draft queries. Except, they won't actually be able to select from that table yet because Snowflake also requires that users have the ability to use warehouses for the compute as well. We'll additionally grant the following:

```
GRANT USAGE ON WAREHOUSE HR_WAREHOUSE TO ROLE SNOWFLAKE_USER;
```

Now that the user has access to a warehouse, they can start running queries and receiving data. **To recap, we must grant access to the object, to all of the containers in the object, and to a warehouse that will do the heavy lifting involved in executing a query.**

51

Revoking Privileges

Revoking privileges is slightly less complicated than granting privileges. When we grant privileges, we're concerned with the hierarchy and with making sure that we grant a privilege at every point in the hierarchy so that the lower-level grants can take effect. When we revoke privileges, we don't need to perform as many revoke statements as we initially did with grants to prevent an asset from being accessible. This is because if there is a broken link at any point between the database and the data in the table, the user will not be able to access the table. We should, however, be thorough because if we reverse any of the revoked privileges, the unrevoked privileges will still exist. For example, if we granted select on a table, usage on the schema, and usage on the database, revoking usage on the database would prevent a user from seeing the schema or the table. However, if we were to ever grant usage on that database again to the same role, that role would then be able to select from the table again since the previous grants are still in effect. To revoke privileges on an object, we structure it like

```
REVOKE <PRIVILEGE> ON <OBJECT TYPE> <OBJECT NAME> FROM ROLE <ROLE NAME>;
```

We can also string multiple privileges together if they're being revoked from the same object and the same role:

```
REVOKE <PRIVILEGE ONE>, <PRIVILEGE TWO> ON <OBJECT TYPE> <OBJECT NAME> FROM
ROLE <ROLE NAME>;
```

If we wish to remove all of the grants to an object to a role, we do not need to be concerned with the ordering, and we can instead issue a statement like

```
REVOKE ALL ON <OBJECT TYPE> <OBJECT> FROM ROLE <ROLE NAME>;
```

When granting and revoking permissions, we should always be aware of what access levels a role has before the grants so that we know that we are exhaustive in the grants we issue. To validate that we've revoked all the correct grants, we can issue the following query:

```
SHOW GRANTS ON <OBJECT TYPE> <OBJECT NAME>;
```

We can also validate that the privileges have been revoked from the role like

```
SHOW GRANTS TO ROLE <ROLE NAME>;
```

We can always grant or revoke additional privileges, but it is much easier to do everything all at once, and then validate, rather than waiting to hear that the user was not able to access what they needed to access, or even worse, that a user was able to access things they should not have had access to.

Privilege Constraints

When we grant access to objects in a database, we need to make sure we grant access to all of the hierarchy above the object; however, we also need to make sure that all of the correct privileges are granted for the task a user may need to do. This is because certain privileges have constraints with other privileges. When we want to update data in a table, that requires us to select from that table, and selectively update certain fields. This means that just granting UPDATE on a table isn't enough, the role must also be granted SELECT.

In addition to constraints between different privileges, there are also constraints between certain privileges and certain roles. For example, the privilege MANAGE GRANTS must be granted by the SYSADMIN role or higher.

Permission Modifiers

In addition to the different types of privileges, there are a few concepts that allow us to expand the reach of a single grant. These modifiers should be used with care since they can affect more than one object at a time, and they can be trickier to reverse. Additionally, there is often a cleaner way of approaching the problem. That being said, these modifiers are very useful and are worth understanding.

All

Distinct from the ALL privilege, the ALL argument allows us to grant a privilege to all existing objects of a specific type within an object. To grant usage to all existing schemas in a database, we can use

```
GRANT USAGE ON ALL SCHEMAS IN DATABASE <DATABASE NAME>;
```

The all argument is very powerful because it allows us to affect many objects at a given time; however, it should be used with caution because it is tricky to reverse. If only one role has USAGE on one schema within a database, and that role is then granted USAGE on all schemas in a database, restoring the state of grants back to the original requires work. We would need to identify the grants that have a timestamp before the timestamp of the ALL grant, and reverse only those received before the grant. This is because granting ALL is the same as granting all of the privileges except OWNERSHIP.

Future Grants

Future grants allow us to grant a privilege to all future objects of a specific type created within the object we're granting access at. The most common example is granting a privilege to all future tables or views within a schema. This would look like

```
GRANT SELECT ON ALL FUTURE TABLES IN SCHEMA <SCHEMA NAME>;
```

The preceding code in combination with ALL allows us to create a policy for all existing and new objects of a certain type in a schema or database. This is as close to a schema-level policy as we will get for these objects.

To view all future grants within an object, we can query like

```
SHOW FUTURE GRANTS ON <OBJECT TYPE> <OBJECT NAME>;
```

Since these future grants are not granted against a specific object, and instead against an object that is to be created, we must specifically query for future grants. This is an important note to remember in the case that grants are queried and logged to an external logging system like Splunk.

It is also important to keep in mind that future grants cannot be used on all object types. Masking policies, row access policies, data sharing, and data replication do not support future grants.

Additionally, future grants do not affect tables that are renamed or swapped. Let's walk through an example of swapping two tables at the same time as we use future grants. Here we're going to grant PRIVILEGE 1 on future tables in this schema to a role. Then we're going to create a test table. When we check the grants to TEST_ONE, we will see that the role has PRIVILEGE 1 on it. We can visualize the state of the schema in Figure 4-3.

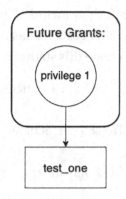

Figure 4-3. *The state of our schema after one future grant and creation of the table test_one*

The state shown in Figure 4-3 will take effect after running the following SQL:

```
GRANT <PRIVILEGE 1> ON FUTURE TABLES IN SCHEMA <SCHEMA NAME> TO ROLE
<ROLE NAME>;
CREATE TABLE TEST_ONE AS (SELECT 1 AS NUM);
```

Next, we're going to revoke PRIVILEGE 1 and issue another future grant for PRIVILEGE 2. Then we're going to create a second table, TEST_TWO. We can first visualize our changes in Figure 4-4.

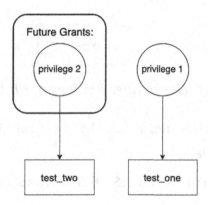

Figure 4-4. *The state of our schema after revoking our future grant, creating a new future grant, and creating an additional table*

55

As we can see, PRIVILEGE 1 is no longer included in future grants; however, it is still applied to our table. We can see that PRIVILEGE 2 has been applied to our newly created table TEST_TWO. The SQL needed to create this state change is as follows:

```
REVOKE <PRIVILEGE 1> ON FUTURE TABLES IN SCHEMA <SCHEMA NAME> FROM ROLE
<ROLE NAME>;
GRANT <PRIVILEGE 2> ON FUTURE TABLES IN SCHEMA <SCHEMA NAME> TO ROLE
<ROLE NAME>;
CREATE TABLE TEST_TWO AS (SELECT 2 AS NUM);
```

Since we revoked PRIVILEGE 1 and granted a second future grant, TEST_TWO will have only PRIVILEGE 2 granted to the role. Next, we're going to grant PRIVILEGE 3 on all future tables in the schema to that role, and revoke our future grant of PRIVILEGE 2. We can again visualize our state in Figure 4-5.

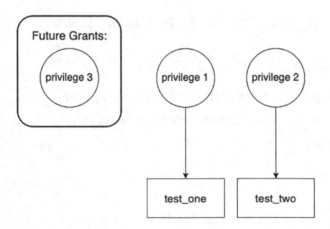

Figure 4-5. *The state of our schema after revoking our future grant*

We can see how our tables are not affected by this change in future grants. To change these grants, run the following:

```
REVOKE <PRIVILEGE 2> ON FUTURE TABLES IN SCHEMA <SCHEMA NAME> FROM ROLE
<ROLE NAME>;
GRANT <PRIVILEGE 3> ON FUTURE TABLES IN SCHEMA <SCHEMA NAME> TO ROLE
<ROLE NAME>;
```

Since these are future grants, they do not affect the two existing tables we created in the preceding snippet. Those two tables still only have PRIVILEGE 1, and PRIVILEGE 2 granted on them, respectively. Now we're going to swap the tables. We can visualize this swap in Figure 4-6.

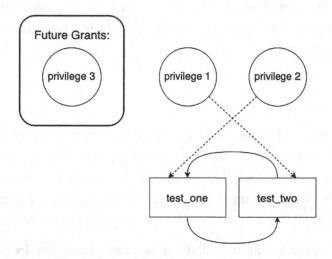

Figure 4-6. *The state of our schema after swapping our two tables*

As we can see in the dashed lines, our grants follow our tables. To swap our tables, we run

```
ALTER TABLE TEST_ONE SWAP WITH TEST_TWO;
```

At this point, we will see that when we swap TEST and replace it with TEST_TWO, the permissions on TEST_TWO carry over, and the future grant of PRIVILEGE 3 does not take effect.

As you can see, future grants require considerably more thought than traditional grants. This means we should be careful with where we use them, and only use them if we think they're necessary.

With Grant Option

Specifying the with grant option in any grant will allow the grantee role to in turn grant that privilege to whomever they would like. This is great in a situation where we want the recipients to be owners who are empowered to allocate access as they see fit. However, it is important to note that this may not be ideal as it leaves an administrator with less control over the Snowflake instance, and it could potentially create a mess.

Working with Permissions

To more easily manage the sheer number of different privileges that exist in Snowflake, we should group them into a few types of roles. This allows us to scale our permissioning in Snowflake and allows us to standardize the way we allocate permissions. To begin, we will group privileges into read, write, and dataset admin roles, since those are fairly straightforward conceptually.

Read

When we group privileges together to create a read role, we want to make it so that datasets are available for viewing, but not for modifying, deleting, creating. Essentially, we want to design a read role such if we were to grant this role to a user, we could guarantee that after a week, the database would be in the exact same state as before the user was granted access. For the read role, we should grant the following privileges:

- SELECT on tables, external tables, views, and materialized views
- USAGE on databases, schemas, and warehouses
- REFERENCES on tables, external tables, views, and materialized views
- OPERATE on warehouses but not on tasks

Write

Write privileges allow a user to create, add, modify, and delete data in schema objects. In some cases, you may want to design a write role such that a write role only has the ability to add and modify data, but not the ability to create or delete schema objects. Additionally, a write role must be granted all of the same privileges as the read role in addition to the write privileges, this is enforced by Snowflake. A role cannot have the ability to modify data in a table without the ability to select from the table. For a write role that only allows the addition and modification of data but not schema objects, I would include the following privileges:

- DELETE rows from tables
- INSERT on tables

- UPDATE on tables

- TRUNCATE on tables

For a write role that has the ability to create and delete schema objects, I would additionally grant these privileges:

- CREATE tags, tables, external tables, views, and materialized views

- APPLY TAG

You may additionally want to include the ability to create and use column and role access policies if the user creating these objects might want to secure those objects. It could be argued that this belongs with an admin role, but I'd argue that there are potential use cases for this to be part of a write role for business intelligence analysts creating sensitive assets. If that is a good fit for your organization, additionally include the following:

- CREATE masking policies and row access policies

- APPLY masking policies and row access policies

Dataset Admin

The dataset admin role is an interesting role because the administrator of a dataset may be different from a general database admin. In the case of team and functional schemas and datasets, which we will visit in the upcoming chapters, a dataset administrator role is ideal. It allows the owner of the dataset to allocate access to whom they see fit, and to truly own their schema or database. If the dataset granularity is at the schema level, then the schema or database level grants below should pertain to the schema, otherwise database. Just like the read role, the dataset administrator should have all of the access of the preceding roles. We can visualize this in Figure 4-7.

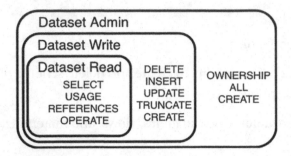

Figure 4-7. *The dataset roles exist in a hierarchy where each level includes the privileges below*

Later in the book, we will cover role inheritance, which makes hierarchies like this even easier. The dataset administrator role should have the following privileges:

- OWNERSHIP on schema or database

- ALL on schema or database

- CREATE stage, file format, sequence, pipe, stream, task, and procedure

- OPERATE task

System Admin

A System Admin, or sysadmin, manages the security, upkeep, and resources of a database, but does not interact with data. This role might be used by a sysadmin on your team but might also be used by a service account through a tool. In any situation, we need to allocate system level privileges to this role. While, unlike the previous roles, a sysadmin may not need to include the privileges granted to the other roles, the recommended best practice is that all roles within Snowflake are granted to the sysadmin role using role inheritance, which will be discussed in a later chapter. In addition to access to all Snowflake roles, a sysadmin should have the following privileges:

- MANAGE GRANTS

- CREATE databases, warehouses, network policies, users, and roles

- ATTACH POLICY

Account Admin

An account administrator manages everything at the global account level. This is our highest-level administrator who can do anything and everything within your Snowflake database. As a result, they should have all of the permissions mentioned previously, in addition to the following:

- CREATE data exchange listings, integrations, shares, accounts

- EXECUTE tasks

- MONITOR EXECUTION

- IMPORT shares

- MONITOR USAGE

Other Specialized Roles

The categories of roles mentioned previously are by no means expansive, but they're an illustration of some common ways to group privileges in a meaningful way. It's likely that your organization may need to create more specialized roles for very particular purposes. It is reasonable to think that a scheduled job that inserts data into a few tables may have its own special role that only allows it to select and insert data into a table in addition to usage on the database, schema, stages, and file formats. It's also reasonable that your organization may need to have a role with very limited constraints for an external SaaS tool. In that case, the purpose of the tool would guide the privileges granted. Another common use case is a resource monitoring role used to monitor things like spend per team.

No matter what groupings your organization decides to implement, the biggest takeaway here is to group these privileges together and to write a tool to generate the SQL.

Key Takeaways

These are the main privileges to consider when designing and implementing role-based access control on your account. Always check the latest version of the documentation on Snowflake's website, as they are continuously adding new features and expanding existing ones.

- The fundamental access control types are read, write, and admin.

- Snowflake has almost 20 types of privileges.

- We can reduce complexity by grouping privileges.

- Create a script to generate each group's grant SQL to be consistent.

PART II

Creating Roles

CHAPTER 5

Functional Roles: What a Person Does

In the last chapter, we covered how we can assign privileges to roles, and how we can create different types of roles based on the privileges they hold. The next step is to understand when to create new roles. Creating roles is not always straightforward. There are many gray areas where it is not clear whether to create a new role or re-use an existing one. To create clarity in the ambiguity, we can use a philosophy or paradigm to guide us.

In this chapter, we will cover how we can create and manage roles based on what a person does for work. Before we get into the who, why, what, and how around all the ways we can think about functional roles, we're going to formally define Snowflake roles.

What Are Roles in Snowflake?

Snowflake uses roles to manage privilege grants to one or more users. We can think of roles as a container for privileges, rather than a container for users. This is because a role is defined by the privileges it has, and a user can have access to multiple roles, and because in Snowflake privileges cannot be directly granted to users, they can only be granted to roles. In the last chapter, we saw how exhaustive the permissions in Snowflake are; as a result, it makes more sense to group privileges into roles rather than granting privileges directly to users. Since we want to reduce complexity, and we want to make our solution scalable, we should create roles based on the datasets in our Snowflake instance, as well as the behavior and characteristics of our users.

Whenever we can identify an access pattern – the same people accessing the same data in the same way – that tells us we have an opportunity to group those people together into a single role. We can create roles based on **dataset**, **job function**, or by **team**.

© Jessica Megan Larson 2022
J. M. Larson, *Snowflake Access Control*, https://doi.org/10.1007/978-1-4842-8038-6_5

To create roles based on dataset or job function, we can use the functional role framework. To create roles based on team, we can create team roles.

How Many Users Should Be in Each Role?

There isn't a simple answer to the number of users per role. In fact, I would expect that there is huge variance between the most populous role and the least populous role in any given Snowflake database. Some roles, like a common role with usage on the main database and most frequently used schemas might be granted to all users in the database. Roles like account admin will only be granted to a very select few. Different organizations may have strikingly large disparities in the number of roles. If our database requires strict access control, then I would expect to see many more roles than you might see in a database with a more laissez-faire policy. In a highly restricted database, you might see a different role for every dataset – in fact, you may see more than one role for each dataset. At the same time, if a database doesn't have sensitive data, it may only have the out-of-the-box roles in place. Again, there are no wrong answers here; every organization is solving for different problems and different use cases.

What Are Functional Roles?

Functional roles are roles based on the function of a user's work. This might mean a characteristic of their job title, like engineer, analyst, or accountant, or it might mean accessing a particular job-related dataset like an internal ticketing system for a customer success team, or a sales tool for the sales team. Functional roles can be boiled down to any role, including users from more than one team accessing the same object or objects in the same way.

Using Functional Roles for Job Function

We can create functional roles around the function of a user's work. When we do this, it is good practice to follow a naming convention that reflects that these are functional roles. For example, we could call an engineering role `functional_engineering`, or simply `snowflake_engineer`. When we do this, we should group all of the privileges that a user needs within this role. Since an engineer may be using their snowflake account to ingest data, it makes sense that they should be able to create types of objects like warehouses, databases, schemas, access policies, and tables. One complication around a role like this is that there is no single "create table" privilege, as we covered in the last chapter. Unlike many SaaS tools that separate permissions based on the ability to do something, Snowflake's permissioning is mainly based on ownership of objects. The ability to create a table is based on the ownership of the schema and database the table will be located in. However, we can use future grants and the all option to grant the ability to create tables in future schemas like

```
GRANT CREATE TABLE ON FUTURE SCHEMAS IN DATABASE <DATABASE NAME> TO ROLE
SNOWFLAKE_ENGINEER;
GRANT CREATE TABLE ON ALL SCHEMAS IN DATABASE <DATABASE NAME> TO ROLE
SNOWFLAKE_ENGINEER;
```

A grant like this allows us to approximate a "create table" permission, though it will not be exhaustive. In the previous chapter, we covered future grants and how they can be trickier to deal with than traditional grants, so that should be kept in mind when creating roles with access patterns like this.

Using Functional Roles for a Dataset

We can also create functional roles based on a dataset a user needs rather than an individual's tasks. The function of a sales operations analyst is to equip the sales team with tools to help them do their job. In this way, dataset-based roles are still functional. Many datasets work well with this model because the data is needed for multiple different teams. We might see this work well with a sales dataset. In this example, we may have an external SaaS tool that we pull data from called Sales Tool; we will create a role like sales_tool_user who has full access to that dataset. This role may have ownership on all objects populated by the pipeline from Sales Tool. Since this role has ownership privileges on this dataset, any downstream table or view should be created by this role in the same location so that everyone has access.

A dataset could also have multiple functional roles. In the last chapter, we covered the different groupings for privileges by function, so we can apply those fundamentals here. It is likely that we have one group of people importing data from Sales Tool, we have a different group of people consuming that data to create downstream tables and views from it, and we have another group consuming the product of that data in a clean, presentable form. In this situation we might want to create three separate roles, `sales_tool_admin`, `sales_tool_write`, and `sales_tool_read` respectively. In this situation, it is entirely possible for a role to only be granted to a single individual. However, using this method to break up roles is safer, and truer to the idea of functional roles.

Why Should I Use Functional Roles?

There are a few different reasons why an organization might benefit from the use of functional roles:

- They might reduce the overall number of roles created.

- They may reduce the overall cost associated with the management of roles.

- They help create clear boundaries at which roles should be created.

- They can be more flexible than team roles,

- They result in fewer changes following organizational structure changes.

Remember, we are using roles to reduce complexity so that we can manage access control at scale, so everything you choose to do with your Snowflake instance should be based on the problems your organization is solving.

Another advantage of functional roles, is avoiding data duplication. When data lives in individual team silos, it can be difficult or impossible for someone outside of that team to surface data assets they may need. In some cases, even if they do surface that data, it may not matter because they are not on the same team, and teams may need to safeguard other assets they have comingled with that data. This might lead to the data being imported into Snowflake more than once, which not only means more overhead and waste, but it also means a higher storage cost. If we know in advance that a particular dataset is used by multiple teams, when we create the role or roles for it, we can allocate them to all of the users who will need access regardless of the team. This maximizes the number of users per role, while at the same time reducing redundancy.

If we know that we're creating one or more roles in connection with a particular dataset, we can name the roles and the schema or database the dataset resides in using the same naming conventions. If we have a schema with ticketing data used to track customers requesting help, we may name that schema the same name as the tool.

Let's say that schema is named `help_tickets`. We might create read, write, and admin roles for that dataset so that we have `help_tickets_read`, `help_tickets_write`, and `help_tickets_admin`. Since they have the same names, it is easy to verify that only the correct roles have access to the data, and it is easy to manage access requests. There isn't any confusion around which roles are for which dataset. If a user asks for access to the `help_tickets` schema and they're not creating any downstream objects, we can immediately give them `help_tickets_read` without having to look up the particular grants to that dataset.

Functional roles help identify where and when roles should be created. Adding Role-Based Access Control (RBAC) to a database that has been an everything-goes system can be intimidating, where does one start? We can start by identifying a dataset, or a group of datasets, that share a function. It is much easier to create roles based on the different datasets than it is to try to separate everyone out into different teams, and then decide which teams have access to what.

Since we loosely define functional roles as being access patterns that span more than one team, it creates a very clear boundary at which we should create a new functional role. Whenever we find ourselves in a situation where a dataset, schema, or any other snowflake asset is used by more than one team, we can create a new functional role.

Finally, there is more flexibility in functional roles than with team roles. Not everyone sits comfortably on a neatly defined team. Not everyone's team handles the same data. How do we break up a team of analysts who all work on different datasets? This is where functional datasets shine, create roles based on what is being accessed and how, rather than based on the reporting structure of the company.

How Do I Use Functional Roles?

Since functional roles are so versatile, we need to balance that out with consistent standards for role creation and use. We need to be consistent with naming so that there is no ambiguity with what a role should have access to. We need to grant these roles in such a way that it is scalable, and easy for users to use. We also need to take steps to make sure that there are expectations and boundaries for users so that we can facilitate collaboration without inciting disagreements between teams.

Earlier in the chapter, we covered different naming conventions for functional roles. In those examples, we made the decision to make the names very concise and descriptive. Roles can include up to 255 characters, more than enough to be descriptive about the type of role, the purpose of the role, and the access granted to it. If a functional role is intended to be used with a particular dataset, make the name of that role the same as the name of the dataset. As we dove into earlier, we can further break this access down into read, write, admin roles, or even special roles like load, or log.

When we name roles, we should append the access type to the name of the role, like `dataset_load`. We can even add on more detail at the end of the name of the role. For example, we could have a role created for the worker loading data into a dataset that doesn't have read access, something like `dataset_load_no_read`. I try to adhere to the following naming conventions: **start general then get more specific, and don't pluralize until the end, if necessary.** When we add detail, we add detail at the end. If we have a dataset that's used by software engineers, but we want a role for managers in engineering for that dataset, we can name the former `dataset_engineer`, and the more specific role, `dataset_engineer_manager`. We also don't want to pluralize until the end, and we only want to pluralize when it makes sense to. For example, we wouldn't want to see `dataset_analysts_managers` and might replace it with `dataset_analyst_managers` or `dataset_analyst_manager`. In this situation, pluralizing manager doesn't add any value, so it's best to do without it. The main reason I like to stick with these two conventions is that it makes searching through text easier. If we want to see a list of all of the roles associated with a dataset, we can easily search for that. Searching for roles in Snowflake is easy, we can query roles directly.

```
SHOW ROLES LIKE '%dataset_engineer%';
```

This would show us both the `dataset_engineer` and `dataset_engineer_manager`. Searching through code in a repository is not nearly as easy, but if we are consistent with our naming conventions, we can easily locate the instances of these roles.

Since we are breaking up roles with more granularity than we might typically have, users will almost certainly have multiple roles at their disposal. This can be difficult to handle at their end, but there are steps we can take to simplify that experience. We should use role inheritance to create a hierarchy of roles, this means that a role would be composed of other roles. We will dive deep into creating these hierarchies later in the book. We can also use secondary roles, which will also be covered later.

To make sure that our functional roles and access patterns are scalable and work well between multiple teams, we need to put some guardrails in place. This means creating expectations for users and delineating responsibilities clearly. We can separate out our datasets into multiple schemas, one for raw data straight from the source, and one or more schemas for transformed data. By directly dumping raw data from the source, we're not only adhering to best practices when it comes to ELT (extract, load, transform), we're also eliminating the possibilities for disagreements and friction over the structure of these source tables. Users should only have read access to the raw database, they should not be creating or modifying any shared assets. We can create a communal transformed data schema for all teams, or we can prescribe that each team does transformation within their own schema.

Who Owns a Functional Role?

Roles do not exist in a vacuum, and neither does a Snowflake instance. These exist within complex and unique organizations with many different teams and users. With team roles or datasets, there is a clear decision maker for changes to a role or dataset. Since a functional role spans multiple teams, that clarity no longer exists. If two separate teams use a dataset, who decides what tables get pulled in? Who decides who gets access to the data and who doesn't? Who decides on golden or certified tables? As with most things, there isn't a single answer to these questions. Ultimately, it is up to your organization how to handle these relationships.

For teams that sit next to each other in the organizational hierarchy, the second-level manager above both of these teams could be a good owner; however, they could also be too removed from the boots on the ground, making them a poor owner of this dataset and role. Either one of those team's managers could be a good choice, though that means that one team could be at a disadvantage. Both of those team's managers could be owners; however, that could still cause disagreements, and doesn't necessarily scale when there are many more teams that use the same functional role.

Another solution is to have a centralized data team own these functional roles. One advantage to this is that all functional roles have the same owner, allowing for the team to create comprehensive policies and procedures for handling access control to these assets. Scaling back the role to read-only on the raw data also makes administering these roles easier because it limits the number of decisions that need to be made around the role. For particularly sensitive datasets, a central data privacy team could make the call on which users should be allowed access and which should not.

At the end of the day, there needs to be a clear owner of each role that can make decisions about access and about the datasets owned by this role.

Key Takeaways

Functional roles come with plenty of advantages, as we covered in most of this chapter. It is important to note that functional roles also come with some disadvantages as well, namely, that there is no clear owner of these roles and assets, which needs to be mitigated before it can create disorganization.

- Snowflake organizes all access control into roles, which can be granted to individual users.

- Functional Roles relate to the function of one's work, rather than the team they're on.

- Functional roles reduce complexity and duplication with datasets shared by more than one team.

- Complexity for end users can be reduced through strategies like role inheritance and secondary roles.

- Naming conventions should be created and used consistently to reduce confusion.

- Functional roles rarely have a clear owner, so decision-making is more difficult.

CHAPTER 6

Team Roles: Who a Person Is

In the last chapter, we covered functional roles. That is, roles that are granted to members of one or more teams. These roles are flexible and are ideal for organizations that collaborate across teams in a cross-functional manner. In this chapter, we're going to introduce the concept of team roles. We're going to explore the types of organizations best suited for team roles, and how team roles and functional roles relate to one another. Most organizations utilize a combination of team and functional roles because the flexibility of functional roles and the structure of team roles allow an organization to cover most use cases.

What Are Team Roles?

Team roles are roles in Snowflake granted to all of the members of a particular team based on the organizational hierarchy. When we use the term team, we're referring to a group of people who work together and have the same manager. This could be a team that includes one or more sub-teams, or it could be a small stand-alone team. Certain organizational structures are more well-suited for team roles; however, all organizations can benefit from them. In Figure 6-1, we can see the relationship between team roles and functional roles.

© Jessica Megan Larson 2022
J. M. Larson, *Snowflake Access Control*, https://doi.org/10.1007/978-1-4842-8038-6_6

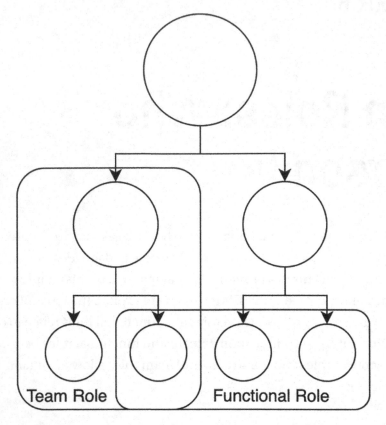

Figure 6-1. *Team roles include members of one team, whereas functional roles can include members of more than one team*

Roles can be created for teams with or without sub-teams. For teams with sub-teams, multiple roles can be created so that a role exists at each point in the hierarchy. For example, if there were a team focused on sales in the Western United States and a team focused on sales in the Eastern United States and they roll up to the director of US sales, then we could create the following team roles: `sales_us_west`, `sales_us_east`, `sales_us`. Alternatively, roles can be selectively created for particular teams rather than all groups of teams. Using the same example, we could create simply `sales_us_west`, and `sales_us_east` if we didn't have a need for `sales_us`. Team roles can be created with some flexibility; there is no mandate to create every single team role that could exist based on the structure of an organization.

Team roles work best for organizations with well-defined hierarchies that do not work cross-functionally and that remain relatively stable over time. An example of this would be an organization that works off of the pod model, where each team works on a particular part of the product. In this model, each team has everyone they need to get the work done, meaning one team could have multiple engineers, a product manager, an engineering manager, a data analyst, and a designer. In this example, everyone having the same role in the database might work particularly well because they do not rely on a centralized data team, the analyst is embedded in their team.

Another clear indication that team roles are right for your organization is if there are teams that use an entirely separate set of data from the rest of the company, or teams that have one or more datasets that are not used by other teams. This may be engineering data around a particular part of the product that can be carefully separated from the rest of the engineering data. This may be particularly true if your organization is relatively small.

Why Use Team Roles?

Team roles help simplify access control by essentially matching the team hierarchy already present within an organization. Team roles are easy to implement, a simple script can be used to create and allocate access to match the employee records. They're also very straightforward, a user needs a team's role if they're on that team. Like we mentioned in the previous section, since we primarily work in teams, they fit many different access patterns.

Implementing team roles can be really easy since most organizations already have an existing hierarchy of team members in software somewhere, whether it is in benefits software, or tax software, or even some home-grown program. A data engineer should be able to whip up a script that pulls that hierarchy from that source system and transforms it into SQL to create and allocate roles. An identity management tool that provisions users and manages employee access lifecycles would also work well with this since we can leverage the System for Cross Domain Identity Management (SCIM) integration with Snowflake. A default role can be specified using the default role parameter during the user creation process. Since organizations already have a need to put users into team groupings, we can leverage existing solutions.

The rules for creating and allocating team roles are pretty straight forward, in general, as long as someone is on a team, they should have access to the role. Organizations can decide whether or not they want to allow team leads or team managers to let other users have access to their team role. However, deciding which assets a team role should have access to is less straightforward. Access requests for role membership are simplified by team roles as well; a team can decide whether or not they want to share their role with stakeholders they work closely with. If we manage team roles using a script like we described earlier, then we can decide to prevent teams from adding additional stakeholders to their role.

Members of the same team are likely to have similar access patterns. It's common for everyone on a team to need the same type of access to much of the same data due to the fact that they're working on similar projects. Because of this, team roles can be a great fit for certain organizations and solve many of the problems they face.

How Do I Use Team Roles?

Unlike functional roles that we covered in the last chapter, team roles typically do not require being broken down by privilege type to as many different types of roles. We discussed breaking up a functional role into read, write, and admin, but that isn't as necessary for team roles. Typically, every member of a team will receive the same role. Naming conventions for team roles should be established and followed for the same reasons we outlined in the last chapter; it simplifies access requests. We can also create team sandboxes using these team roles, which can allow teams to work more swiftly. When using team roles, we should be aware of what datasets exist where because usage patterns with team roles can result in data duplication, so it is good to get ahead of it before hundreds of nearly identical tables exist.

When we work on teams, typically the members of a team have the same or similar job function, and report to the same manager. As a result, it usually doesn't make sense to break up team roles into read, write, and admin. We may not need to break them up at all, or we may simply create write and admin so that only one member of the team has the ability to delete the Snowflake objects. Teams that work under a pod model might still benefit from the read, write, and admin distinction.

Naming conventions for team roles are really important, especially for larger organizations with more employees. In a large organization, there may not be any single accounting team, there may be 12 that all focus on different things under the accounting umbrella. For this reason, I recommend using the same general-to-specific convention, so instead of `revenue_accounting`, I might instead create `finance_accounting_revenue`, where the name of the role includes the entire hierarchy above the team. Small organizations don't necessarily need to be this strict with their naming conventions, if there are 100 people at a company, there's probably only one accounting team.

Additionally, we can use team roles to create team-level sandboxes to allow teams to work together more efficiently and quickly. In a situation like this, members of a team might create downstream tables from the raw data that's stored in a different schema. There may be no clear-cut location for these tables to go, so a team sandbox schema is ideal in this situation. However, we need to use caution when we create sandboxes like these because they create data silos.

We will cover downstream services later in this book, but let's touch on this really briefly. One great way to prevent data duplication in these silos is to utilize a data discovery tool that surfaces metadata around all datasets, including locations most users won't have access to. These tools typically include table names, field names, and lineage for these tables and fields so that it is clear where the data is coming from. This allows other potential users of the data to spot where it is and ask for access, at which point the resource should be moved out of the team sandbox and into a location with higher visibility.

Who Owns Team Roles?

With a higher focus on data governance now than ever, there is an additional push for every object to have a clear owner. In the case of team roles, ownership is a lot clearer cut than with functional roles. Since teams typically follow a similar hierarchical structure that rolls up to one or more managers or leads, there is a natural owner or owners that can be utilized. If a team has a tech lead in addition to a manager, it is up to the team whether they want the lead or the manager to own the role. Since a tech lead is typically more in the weeds with the actual work the team is doing, that is more preferred than a manager.

It's also possible that the best solution is for a particular analyst, data engineer, or data scientist on the team to own the role. In the case of organizations that work in pods, if there is one person on a team who is using Snowflake more than others on the team, it might make the most sense for that person to handle all the administrative tasks in Snowflake, including owning the team role. Since this user may be the only one creating objects in Snowflake, it makes sense for them to own the entire role.

A centralized data team can also own team roles. In the situation where team roles are automatically created and updated using a pipeline that pulls the organization's employee data, it makes more sense for a centralized data team to own all team roles. In this case, the tool and the pipeline really own the roles, since that is the logic that creates and populates roles; however, since we need a human owner for all objects, the owner of that particular pipeline should own the roles.

When Should I Use Team Roles and When Should I Use Functional Roles?

We should use team roles when we have teams of users with similar access patterns, which exist within most organizations. I recommend using team roles initially, they're really easy to set up and they are easy to work with. By allocating team roles to every existing Snowflake user in our organization, every user belongs to a particular team role, and hopefully has a team sandbox to use for analysis.

When access to a dataset or object spans more than one team, or may span more than one team in the future, we should instead create functional roles for that object. It is much easier to grant a functional role to additional users at a later date than it is to uproot that dataset from a team schema and move into a functional framework.

How Can I Use Team Roles and Functional Roles Together?

Team roles and functional roles complement each other really well and work best when used together. As I mentioned in the previous section, I recommend that every user in the organization has access to a team role. We can build on top of this by adding functional roles into the mix. In the last chapter, we briefly addressed using role inheritance with functional roles; we can also use role inheritance with a combination of team and functional roles. If a team uses multiple datasets, using functional roles in combination with team roles simplifies the requirements for onboarding a new team member. Additionally, we can leverage both of these concepts when working with shared assets.

When using role inheritance for functional and team roles, we need to understand whether or not the entire team needs access to a functional role, or whether it should just be granted to a single user on the team. If the entire team needs access to a functional role, then we can grant the functional role to the team role using role inheritance. If a team needs access to many shared functional roles, then each of those can be granted to the team role as well. By doing this, we greatly reduce the work required to either add or remove a member of the team, or to add or remove a team's access to a shared dataset. To remove access, simply revoke the team or functional role. To add access, simply grant the team or functional role. When we grant access to roles in a role hierarchy, we should take care to revoke redundant roles; that is – if users are directly granted a role that is also inherited by an available role for the user, we should revoke the directly granted role.

Using role inheritance is significantly less work than individually granting every single role, each modification requires only one line of SQL. Since it is more systematic and concise than granting to each user, it is also safer; when granting or revoking multiple things at a time, it is easier for something to fall through the cracks. An added bonus is that if the team roles are created and managed by the source of truth for employee status and team membership, it requires zero work in Snowflake on the part of the administrator of Snowflake, since the pipeline should propagate changes to the upstream data.

As we covered earlier in this chapter, team sandboxes work really well with functional datasets. In a situation where a functional role and schema `engineering` exist, which contains data like `user_activity_logs`, a team that works on the user profile experience, `engineering_user_profile`, might pull all relevant data like

```
SELECT * FROM engineering.user_activity_logs WHERE page_type = 'profile';
```

They might pull this into a table or view and store it in the `engineering_user_profile` schema so that it contains all of the same columns as the original table but filtered, so that the result is only data relevant to that particular team. Patterns like this respect the abstraction barriers posed by the functional and team roles.

Key Takeaways

As we covered in this chapter, most organizations can benefit from the use of team roles and should employ a role-based access control strategy utilizing them. Unless your organization frequently does widespread reorganizations, team roles can greatly simplify the work that goes into maintaining access control. Strategies we discussed in this chapter can facilitate the use of team roles as well.

- Team roles work great for teams where everyone on the team works on the same datasets.

- Team roles used in conjunction with functional roles can simplify Snowflake administration.

- Team roles need owners; this can be a manager, lead, analyst, or even a centralized data team.

- Consistent naming conventions are more important for larger organizations, but are always encouraged.

- Sandbox schemas for each team can speed up development, but they can also lead to data duplication.

In the next chapter, we will dive into how users can start to work with roles in Snowflake.

CHAPTER 7

Assuming a Primary Role

We've covered the different role paradigms, whether functional or team, and we've addressed the different privileges Snowflake provides. We've discussed how we can draft roles so that they are manageable and scalable. Now we're going to cover how we actually use these roles in practice.

Snowflake is based on role-based access control; all objects are owned by roles rather than by individual users. Privileges are granted to roles rather than directly to users. Users can switch between the different roles granted to them, assuming one primary role at a time. In this chapter, we're going to cover what primary roles are, what default roles are, and how to use primary roles.

What Is a Primary Role?

It is very rare for a Snowflake user to be assigned only one role. It is much more common for a user to have many roles with a range of permissions at their disposal. To utilize these privileges, we must select and assume a single role at a time. This is because some of the crucial privileges can only be granted to one role per object. For example, ownership of an object can only be granted to one role. When we assume a primary role, we are telling Snowflake to ignore all of the other roles that we have access to, and to only let us use the privileges granted to this one particular role.

We can easily switch between different roles within a session; however, we cannot utilize multiple primary roles in a single query or transaction. This is because a transaction is the smallest unit of activity in Snowflake, and that activity needs to be logged into system tables and views and attributed to a single role. Whichever role is our primary role will be the role used for logging purposes. Additionally, there is no mechanism with which to switch primary roles in the middle of a query.

© Jessica Megan Larson 2022
J. M. Larson, *Snowflake Access Control*, https://doi.org/10.1007/978-1-4842-8038-6_7

Default Roles

Every user has a default role assigned to them at account creation. This is the role that is assumed without a user needing to specify a role. Certain downstream services, including visualization tools like Tableau, do not provide a way to change roles, so some organizations may need to dedicate resources towards assigning the correct default role to all of their users. Administrators can change the default role assignments, and users can update their own default roles.

When a user opens up the Snowflake user interface (UI), connects through a Python connector, or even connects through a third-party application, the default role is the role that is used without needing to specify which one to use. This means that properly assigning the appropriate default role to users can save everyone time. Fortunately, in most of these situations, if the default role is not the most appropriate choice, users can override that and select another role they have at their disposal.

Some downstream tools may not have the ability to specify a role other than the default role. This can pose some issues. In this situation, we need to dedicate resources to assigning default roles at scale so that the default roles match the privileges a user needs when using this tool. This rigidity with tools becomes especially problematic for users that work cross-functionally and need to utilize multiple roles depending on the day or the project they're currently working on. In this situation we have tools we can use to make this work. I've touched on role inheritance a few times already because it is a very powerful tool that allows us to grant a role and its privileges to an entirely separate role. For this reason, we will spend an entire chapter on this later in the book. Additionally, we can utilize secondary roles to solve this, which will be covered extensively in the next chapter.

Default roles are not forever – administrators with the ability to edit user characteristics can update the default role for any user in the organization. This can be done at any point in time and can be done unlimited times. The only constraint is that only one default role can be assigned to each user at any given point in time, and that user must have that role granted to them. To change a user's default role, simply run the following query:

```
ALTER USER <USERNAME> SET DEFAULT_ROLE = <ROLE_NAME>;
```

Individual users can also update their own default roles using the same SQL query. Since this parameter is set at the user level, it will persist across new sessions.

How Do I Assume a Primary Role?

Assuming a primary role is quite simple, and we can do this in a few different ways depending on where and when we choose to change roles. While we have different methods, they all inevitably boil down to a SQL statement that Snowflake executes during an open session or connection. Changing a primary role during a session only lasts until the end of the session. Once a connection is closed, the next session will be opened with the default role.

Using SQL

We can assume a primary role using SQL. This is the most fundamental way of doing this and is what all of the other methods are doing under the hood. To switch to a different primary role during an open session, we can simply run

```
USE ROLE <ROLE_NAME>;
```

When you do this at the beginning of a session, Snowflake will stop associating your activity with your default role, and then start to associate any new activity with the new primary role.

Before we switch roles, we may want to know which role is currently our primary. We can easily do this by running the following SQL:

```
SELECT CURRENT_ROLE();
```

We may also want to know which roles we have available to us, which we can see by running

```
SELECT CURRENT_AVAILABLE_ROLES();
```

This will return a row with an array of all of the roles available to us, the querying users. We can use any of these roles without generating an error.

Using the Snowflake User Interface

There are a few ways we can change our primary role in the Snowflake UI. We can do this using SQL in the worksheets, or we can use one of the dropdown menus to select the role we would like to use. If we're using the classic UI, we have two separate locations we can use to change our primary role. This is because different parts of the UI are populated separately.

Using the classic UI, we have two separate drop-down menus where we can select a primary role. One is on the very top right, which controls the top menu that allows us to browse database objects, account information, worksheets, and query history. We can see this menu in Figure 7-1.

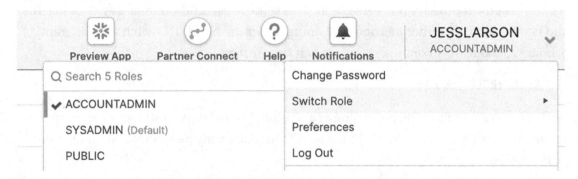

Figure 7-1. *Switching a role in the Snowflake UI*

Worksheets each can have their own primary role. To change the primary role used while executing queries in a worksheet, we can either use SQL, or we can use the drop-down menu within the worksheet pane. This menu is on the right side of the screen, just below the UI's main role selection, and includes other options like the warehouse, database, and schema, as shown in Figure 7-2.

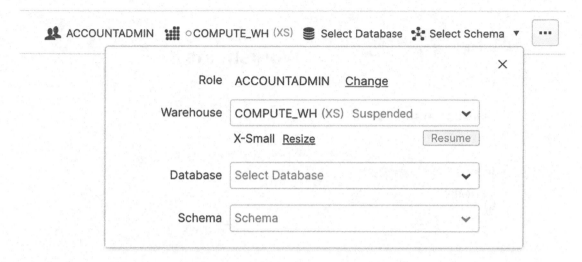

Figure 7-2. *The worksheet menu allows a user to change their primary role on a worksheet*

As we can see in the top left corner of Figure 7-2, our current role is ACCOUNTADMIN. Since we can flip between multiple worksheets while we work in the Snowflake UI to work on different things, it makes sense that each of these worksheets should have its own role.

The new UI makes switching roles a bit easier than the classic one. When we first open the app, the landing page allows us to select a role for the UI. This menu is on the top-left and also controls the information displayed in the UI, as shown in Figure 7-3.

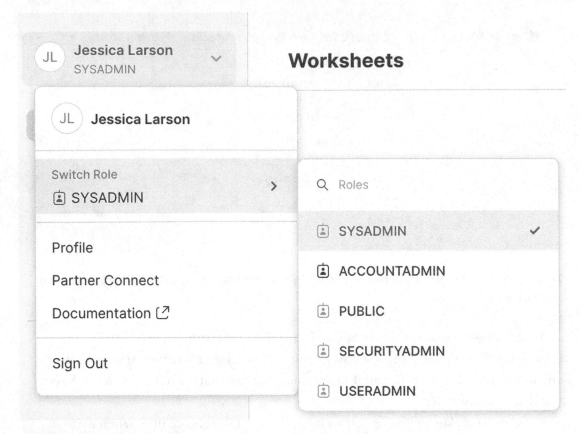

Figure 7-3. *The new UI role menu controls the population of the UI*

When we work on worksheets in Snowsight, the newer version of the UI, this menu is no longer visible, making it a bit more user friendly than the classic UI. We can still easily see our current role, which is SYSADMIN denoted by the check mark, as well as its location on the left side of the drop down. This then gives us the option to change the worksheet primary role using the menu on the top left, as shown in Figure 7-4.

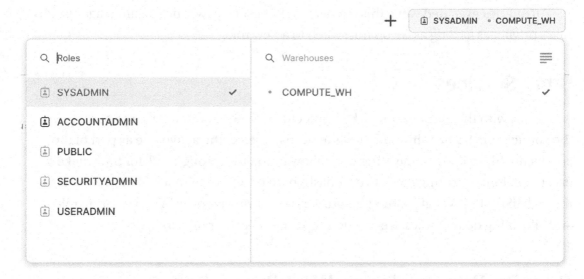

Figure 7-4. *Selecting a worksheet primary role in the new UI*

With all of these methods, switching to a role for the entire UI will fall back to the default role when we log back in. However, for the worksheets, the role will stay tied to the particular worksheet because each worksheet has its own session. It is also important to note that regardless of the UI we're using, when we open a new worksheet, that worksheet will use our default role.

Using Python Connector

When we use the Python connector, we create a connection to Snowflake using our user characteristics or using a service account. We can specify parameters within our connection definition, or we can execute SQL to change our role. To specify the role within connection parameters, we can instantiate a connection like

```
conn = snowflake.connector.connect(
        user=<USER>,
        password=<PASSWORD>,
        account=<ACCOUNT>,
        role=<ROLE>,
        )
```

All queries executed while this connection is still active will be executed against this role's privileges as long as we do not switch to a new role.

Other Services

As we saw with the Python snowflake connector, we can specify the role as part of the connection. Some third-party tools allow us to select the active role as part of the connection definition. Some other tools allow us to pass through SQL for Snowflake to execute before executing the SQL on a dashboard or report, in that situation we can pass through `USE ROLE <ROLE_NAME>;` before we view or run a report. When creating roles and allocating default roles, we should keep our tool constraints in mind.

How Do Primary Roles Work?

When we assume a primary role, we're telling Snowflake to ignore all of the other roles, and to only respect the privileges granted by this particular role until the session is terminated, or until the role is changed. This means that we will only have access to the datasets granted to this role while we use it, and not the other datasets we have been granted through other roles. The current primary role is also the default owner of all objects created during the session.

Changing roles within a Snowflake session is actually overriding a session-level parameter for the current role. When we think of changing roles like this, it makes sense that this switch would not persist across different sessions and would not persist after terminating a session. Every time we open a new session with Snowflake, the session starts with our default role. If we wish for the role we assume to persist across sessions, then we should update our default primary role as explained earlier in the chapter.

Since we can only assume one primary role at a time, we will only have access to the privileges granted to that particular role. This means that we need to draft our roles such that roles have sufficient privileges to do everything necessary to complete tasks. If a user needs to create a downstream table in one schema using data pulled from a different schema, then they need a role that has write access on one schema, and at least read access on the other schema. Let's explore this in the following example. User `USER_1` has two roles granted to them, role `ROLE_A` which has read access on schema `SCHEMA_A`, and role `ROLE_B` which has write access on schema `SCHEMA_B`. User `USER_1` needs to create table TABLE_B in schema SCHEMA_B using data pulled from table TABLE_A in schema SCHEMA_A. We can visualize this in Figure 7-5.

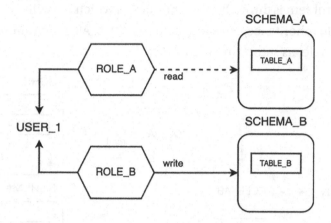

Figure 7-5. *User USER_1 has read access on schema SCHEMA_A and write access on schema SCHEMA_B through role ROLE_A and role ROLE_B, respectively*

Without any other privileges, user USER_1 cannot create table TABLE_B because pulling data from a source and storing it somewhere, even using temporary tables, requires create table or create view privileges in at least one schema. We can solve this problem for user USER_1 in two ways. First, we can create a role ROLE_AB that has read access on schema SCHEMA_A and write access on schema SCHEMA_B, as shown in Figure 7-6.

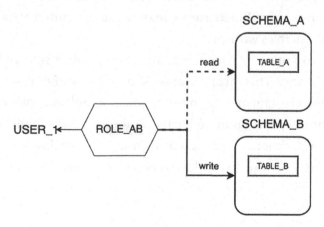

Figure 7-6. *Role ROLE_AB has read access on schema SCHEMA_A and write access on schema SCHEMA_B*

In this situation, user USER_1 can now pull data from schema SCHEMA_A in the same transaction as creating table TABLE_B in schema SCHEMA_B using role ROLE_AB. The other way to solve this problem is through role inheritance, which we will cover in upcoming chapters. Using role inheritance, we can create role ROLE_AB and grant role ROLE_A and role ROLE_B to that role, as shown in Figure 7-7.

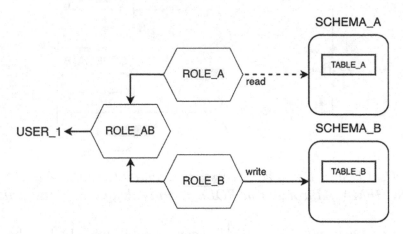

Figure 7-7. *Using role inheritance, role ROLE_AB is granted all privileges from role ROLE_A and role ROLE_B*

Role ROLE_AB when created from scratch and when created using a hierarchy still functions in the same way and affords the user the same privileges. Role inheritance simplifies the process of creating these complex roles because it only requires one grant per role in the hierarchy. Since users can only use one primary role at a time, it is important to draft these roles with care.

The last big piece of primary roles is that the active primary role for the creator of an object is the default owner. This means that we should be careful with which primary role we're using when creating objects in a location that multiple roles have access to. If we're not using future grants on these locations, then we want to make sure that all objects created in this location are created using a shared role that everyone who needs the data has access to. This way no one loses access to important assets.

Key Takeaways

When we assume a primary role, remember that we're selecting a subset of our privileges that we would like to use in our session. Since privileges are granted to roles, rather than users, when we assume a primary role, we can only use the set of privileges granted to our current primary role.

- Users can have many different roles; however, they can only use one primary role at a time.

- Every Snowflake session starts with the user's default role but can be changed.

- Administrators with user privileges can update a user's default role, and the user can update their own.

- Some tools allow us to specify a role as part of the connection, some do not; keep this in mind when creating and allocating roles.

- Changing active roles is actually modifying the session parameters, which do not persist once the connection is terminated.

- The active primary role is the owner of all objects created by a user.

In the next chapter, we will cover how we can use secondary roles to use the superset of our privileges.

CHAPTER 8

Secondary Roles

Secondary Roles are a newer addition to the RBAC system in Snowflake. In the past few chapters, we've run into some limitations with how primary roles work, specifically for users that work cross-functionally and need to assume multiple roles. One of the solutions I mentioned was role inheritance, which we will cover in the next chapter. The other solution I mentioned is secondary roles.

At Pinterest, I was an early adopter of secondary roles. We had Snowflake connected to Tableau in such a way that users needed to authenticate using single sign on (SSO) to Snowflake through Tableau in order to view a dashboard. This forced the dashboard to refresh for each user to guarantee that the information presented in the dashboard was data the user was allowed to access, since our source of truth for RBAC is Snowflake. One issue we ran into was that when creating a dashboard using Tableau Desktop and defining a database connection, analysts would need to either specify a role that everyone viewing a dashboard would need to assume, or they would leave the role parameter blank, forcing all users to use their default role. Unfortunately, this meant one of two things; if we went with the specified role, then every user viewing that dashboard would need access to the same role, and if we went with the default role, then we would need to be very clever with assigning default roles so that every user's default role combined all of the privileges they would need in Tableau. Unfortunately, neither of these solved our problem for cross-functional users without a significant amount of work. We also had two members on our Snowflake Platform team at the time, so we didn't have the resources to dedicate to perfecting the roles assigned to users. Fortunately, our rock star sales engineer brought our attention to Secondary Roles, which was in private preview at the time.

Using Secondary roles allowed us to utilize the default role in the connection definition, while in the background, the combination of all of the user's roles were invoked, allowing them to access all data that any of their roles have access to.

© Jessica Megan Larson 2022
J. M. Larson, *Snowflake Access Control*, https://doi.org/10.1007/978-1-4842-8038-6_8

This solved our problem because then we didn't need to make sure that every single user accessing a dashboard has access to the same role, and we didn't need to spend time and energy assigning the correct default role to every user. It solved the problems we were losing our minds over and allowed us to focus our resources on other problems.

What Are Secondary Roles?

In the previous chapter, we covered primary roles – the main role assumed during a Snowflake database session. Secondary roles are essentially the inverse of primary roles. While you can only invoke a single primary role at a time, secondary roles allow a user to assume all applicable roles at a time, which means they can use all of their privileges at once.

In the last chapter, we identified a situation where a user had read access to a particular schema, and a separate role with write access on a different schema as shown in Figure 8-1.

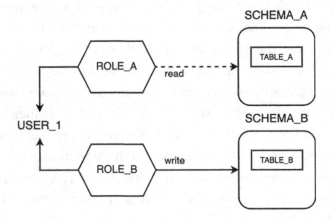

Figure 8-1. *User USER_1 has role ROLE_A providing read access to schema SCHEMA_A, and role ROLE_B providing write access to schema SCHEMA_B*

Because the read role did not allow the user to create a table or view, they would not be able to use that role in order to create a downstream table, and, at the same time, the write role did not provide them read access to the data they needed to consume, so that role could not be used to accomplish the task either. We went through an example of using role inheritance, but we can also solve that problem using secondary roles.

Without using secondary roles, this user would not be able to create a downstream table from table TABLE_A. With secondary roles, the user can utilize the read privileges granted to them through role ROLE_A in the same query as they utilize the write privileges granted to them through role ROLE_B. Since creating assets uses the active primary role for ownership, user USER_1 would need to use role ROLE_B as their primary role to finish this task.

Secondary roles are a game-changer, they allow roles to be used with greater flexibility. In the situation I encountered at Pinterest, it allowed us to maintain the strict RBAC we had in place, while at the same time it freed up resources and allowed us to focus on other tasks rather than dedicating more time to role management.

Why Should I Use Secondary Roles?

Whether your organization uses functional roles or team roles, or both, secondary roles are the perfect glue to put everything together. Organizations that work cross-functionally can benefit the most from secondary roles because cross-functional access patterns typically mean invoking multiple different roles depending on what the particular project calls for. Additionally, secondary roles can solve problems with third-party tools. As we will see in the next section, secondary roles do not require a lot of lift to set up either, and can easily be disabled, so it is worth experimenting with them to see if they're a good fit for your organizational structure.

Secondary roles allow teams to work more efficiently by focusing on what privileges a role has, and what users have access to, rather than spending time making sure that each user has a role that allows them to do everything they need at once. As we saw in the example in the previous section, secondary roles act as a glue, combining the read role's privileges with the write role's privileges to allow a user to complete their task of creating a downstream table using data from a separate schema.

Organizations with complex RBAC systems, or those that collaborate cross-functionally can stand to benefit the most from secondary roles. Since secondary roles stitch permissions together, a user with read access to three different schemas using three different roles, with write access to a different schema using a different role would be able to pull data together from those three source schemas to create a table in the fourth target schema. Analysts who work in patterns like this won't need to think about which role they need to use for which problem, they will only need to pay attention to which role they use to create assets.

Secondary roles solve issues with third party tools that require role specification in the connection definition, or otherwise use the default role. These tools are typically the presentation and visualization layer, mostly used by less technical users. While an analyst may use the Snowflake UI or the Python connector, a business user may only interact with Snowflake through a dashboarding tool. Since these users are typically the least technical out of the Snowflake users, it may not be reasonable for them to need to understand the RBAC system at play. They may be granted multiple roles based on their job and team without knowing which roles they're granted. Using secondary roles with these third-party tools, the end user will have a seamless experience; they will not need to choose a particular role, everything should just work. The beauty of this is that it respects the abstraction barrier of the separation of work responsibilities; a business user shouldn't need technical expertise or savviness unless their job requires that of them.

Most organizations can benefit from secondary roles, though they can be a bit of a black box. They are not as straightforward as primary roles, and logging and understanding which role is used at which time is not always clear. We will explore this later in this chapter.

How Do I Use Secondary Roles?

Enabling secondary roles at an organization requires a few steps. First, secondary roles will need to be enabled for all existing users. Once all existing users have secondary roles enabled, we will need to ensure that new users also have access to this feature. Next, secondary roles will need to be enabled for any integration that supports and will use that functionality. The final thing to keep in mind is that user activity is not logged the same with secondary roles as it is with primary roles.

Enabling Secondary Roles for Users

Each user will need to have secondary roles enabled individually. Currently, secondary roles use every role available for a user, but the roadmap includes being able to select a subset of roles for use. To enable secondary roles for a user, simply run the following query:

```
ALTER USER <USERNAME> SET DEFAULT_SECONDARY_ROLES = ('ALL');
```

As shown in the preceding query, the input to the parameter DEFAULT_SECONDARY_ROLES is a list including all roles. In the future, this is intended to be a customizable list of roles that a user can assume using secondary roles. In that situation, it would be ideal to have a list of roles to exclude from secondary roles. Since this parameter needs to be enabled for all users in an organization, it makes sense to do this programmatically initially. I did this using the Python connector to pull all non-service account users, then generated the SQL, and passed it to the Python connector to run in Snowflake. We could also do this same set of steps using procedures using JavaScript to loop through the users. I prefer working with Python because I use Python more frequently than JavaScript, but both options will do the job. When we draft our script, we want to skip service accounts because those accounts often do heavy lifting, and it may be important to know which roles were invoked by the service account at each step in the pipelining process.

Now that all existing users have secondary roles enabled, we need to make sure that all new users get that functionality. We have two options, depending on how we create Snowflake users. If we use the System for Cross Domain Identity Management (SCIM) integration, we can simply add the parameter DEFAULT_SECONDARY_ROLES = 'ALL' to our user creation parameters. If we manually create users using SQL, then we add that parameter to our SQL query. We can see an example here:

```
ATE USER <USER NAME>
...
DEFAULT_SECONDARY_ROLES = ( 'ALL' );
```

This will automatically provision new accounts with secondary roles enabled.

Enabling Secondary Roles on Integrations

Now that we have ensured that all of our existing and future Snowflake users have secondary roles enabled, we need to enable the third-party applications. This works for a set of third-party tools that are connected to Snowflake through a security integration. In my real-life example at Pinterest, I needed to enable secondary roles on the Tableau security integration. To do this, simply execute the following query:

```
ALTER SECURITY INTEGRATION <INTEGRATION NAME> SET OAUTH_USE_SECONDARY_ROLES
= 'IMPLICIT';
```

This will need to be done for every single security integration that will need to utilize secondary roles. At Pinterest, this meant enabling the feature for both Tableau Desktop and Tableau Online/Tableau Server.

Logging and Ownership

Since secondary roles necessarily mean a user invoking multiple roles at a time, Snowflake doesn't log a user's activity the same when using secondary roles as one might expect. If a user creates a resource while secondary roles are enabled, the resource will belong to the primary role during that session.

Snowflake logging tracks the user's primary role at the time the action takes place. This means that if a user is selecting data from a table that their primary role does not have access to, but one of their roles does have access to, the logging will reflect that the primary role completed this action. This may complicate some auditing depending on your organization's requirements. A check to make sure that only the correct roles are accessing the correct datasets might raise an error when in fact the dataset was not exposed. In Figure 8-2, we can see how this might happen.

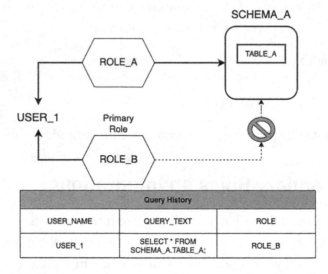

Figure 8-2. *User USER_1 uses primary role ROLE_B to access data in table TABLE_A in schema SCHEMA_A; ROLE_B does not have access but is logged as the role taking this action*

In Figure 8-2, user USER_1 has two roles, role ROLE_A and role ROLE_B, and selects role ROLE_B as their primary role this session. Using secondary roles, they're able to access data from schema SCHEMA_A since role ROLE_A has access. We can see how this would work in the following. Before we enable secondary roles for USER_1, we can assume ROLE_B as our primary role, and then attempt to select from our table.

```
USE ROLE ROLE_B;
SELECT * FROM SCHEMA_A.TABLE_A;
```

Since ROLE_B does not have access to SCHEMA_A and we do not have secondary roles enabled, this query will fail. Now, we're going to enable secondary roles and try again.

```
USE ROLE ROLE_B;
ALTER USER USER_1 SET DEFAULT_SECONDARY_ROLES = ( 'ALL' );
SELECT * FROM SCHEMA_A.TABLE_A;
```

Since we've enabled secondary roles for our user, this query will succeed.

The logs reflect that role ROLE_B completed this action even though role ROLE_B does not have permissions on this schema. In this situation, if your security team is concerned about users accessing datasets they should not have access to, it might be best to set up auditing that checks all of the roles that a user has access to and sends an alert if that user does not have access to an allowed role for a dataset.

Since assets belong to the role that created them, users creating assets will still need to consider which primary role they have set in a session. Let's walk through an example. In this example, we're going to use a revenue_accounting schema, which has a revenue_accounting_analyst role with write access. Our user also has a finance_analyst role that can be used to create tables in the finance schema. We can also assume that secondary roles have been enabled for our user. The finance_analyst role does not have any privileges on the revenue_accounting schema. We can visualize this in Figure 8-3.

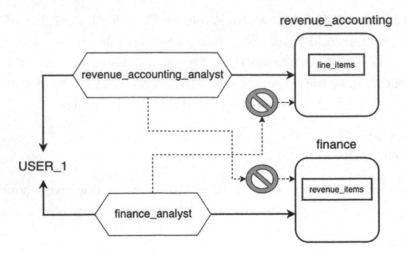

Figure 8-3. *User USER_1 has access to the revenue_accounting_analyst and finance_analyst roles*

We're going to use the finance_analyst role to create a table in the finance schema that pulls data from the line_items table in the revenue_accounting schema.

```
USE ROLE finance_analyst;
CREATE TABLE finance.revenue_items AS (
SELECT *
FROM revenue_accounting.line_items);
```

We need to assume the finance_analyst role to complete this operation because the revenue_accounting role does not have write access on the finance schema. Since we have secondary roles enabled, our user is able to use the privileges from the revenue_ accounting role to select data from the revenue_accounting schema.

Because write access still uses the primary role, we don't need to worry about enforcing the correct ownership of objects. When secondary roles are enabled, Snowflake will force users to assume the primary role with adequate permissions for creation of objects.

Disabling Secondary Roles

In the last section, I mentioned that secondary roles are easy to set up, and that I encourage data teams to experiment with them since they are also quite simple to disable if they aren't the right fit for an organization. To disable secondary roles, we will need to disable them for all users, and disable them on any integration with them enabled. To disable secondary roles for users, run this query for each individual user:

```
ALTER USER <USERNAME> UNSET DEFAULT_SECONDARY_ROLES;
```

This can be done programmatically through the Python connector. The next step is to disable secondary roles for each integration using the feature.

```
ALTER SECURITY INTEGRATION <INTEGRATION_NAME> UNSET OAUTH_USE_
SECONDARY_ROLES;
```

Disabling secondary roles on an account requires very little work, as does enabling them. This makes experimenting with this feature simple, low-effort, and low-risk.

Key Takeaways

Secondary roles are a huge win for both technical and non-technical team members. Snowflake does work in the background using secondary roles that allows us to focus on other things rather than role maintenance and educating our users about how to use roles. The biggest wins for secondary roles come through tools used for visualization by end users. When we use secondary roles, these users need not know what roles are, creating a much more pleasant and seamless experience for those users, allowing them to focus on the core aspects of their job.

- Secondary Roles simplify the RBAC process by combining all available privileges.

- Organizations with many roles or that work cross-functionally benefit the most from secondary roles.

- Write privileges depend on a user's current primary role; users will need to assume a role with write privileges to create assets.

- Secondary Roles may create confusion in logging; the primary role will be logged even if it doesn't have access to the dataset.

- Enabling this feature requires very little work, as does disabling it, making experimenting with secondary roles easy and low-risk.

In the next chapter, we will cover role inheritance, which can provide some of the same benefits as secondary roles.

PART III

Granting Permissions to Roles

CHAPTER 9

Role Inheritance

In the last chapter, we covered secondary roles, which allow users to stitch together privileges granted through different roles. This is really powerful in organizations that work cross-functionally, as well as in organizations with a large, complex RBAC system in place. We can use role inheritance in a similar way to empower our users and allow them to utilize privileges from multiple roles during the same transaction. This also means they will not need to switch between roles as frequently. By using role inheritance, we can maintain abstraction barriers; end users should not need to think about how the RBAC system is structured, they should be able to use Snowflake without thinking about which role they need to use for which data. In this chapter, we're going to explore role inheritance and how we can use it in different ways, depending on how we are using roles at our organization.

What Is Role Inheritance?

Role inheritance is a feature in Snowflake that allows a role to inherit all of the permissions from another role. We can use role inheritance to create a role hierarchy, simplifying access. When we set up role inheritance, we are granting all of the permissions within a role to a new role. This not only grants all of the privileges associated with the role, but it also gives the user access to that particular role, meaning that users can individually assume each role as well as using the role's privileges in a parent role.

When we create a role hierarchy, we have one or more parent roles combined with one or more child roles. The parent roles inherit all of the permissions from the child roles and can additionally have permissions directly granted to them. We can visualize the inherited permissions in Figure 9-1.

© Jessica Megan Larson 2022
J. M. Larson, *Snowflake Access Control*, https://doi.org/10.1007/978-1-4842-8038-6_9

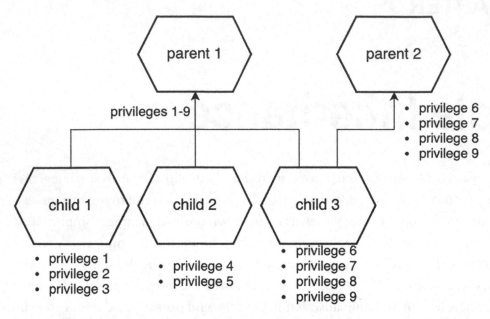

Figure 9-1. *Child role 1 has privileges 1–3, child role 2 has privileges 4 and 5, and child role 3 has privileges 6–9. Parent role 1 inherits all 1–9 privileges, whereas Parent role 2 inherits privileges 6–9*

Since all 1–9 privileges are granted to parent role 1, a user assuming parent role 1 can utilize privilege 2 during the same transaction as utilizing privilege 7. Without using role inheritance or secondary roles, this would not be possible. An admin would need to create a role that encompassed all of the privileges for these to be used at the same time. Additionally, we can see in Figure 9-1 that child role 3 has been granted to more than one parent role.

Role inheritance also grants the role itself to a user granted the parent role. This means that a parent role that inherits three roles like in Figure 9-1, if granted to a user, would grant that user the ability to use four roles in total. We can visualize this in the following figure.

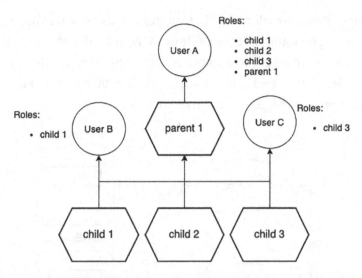

Figure 9-2. *User A is granted parent role 1, which is granted child roles 1–3, giving user A all four roles. User B has child role 1 and user C has child role 3*

In Figure 9-2, we can see that the child roles 1 and 3 are granted directly to users, giving each of those users one role they can assume. Child roles 1 through 3 are granted to parent role 1, which is then granted to user A who now has access to a total of four roles. User A can use child roles 1–3 in the same way as they can use roles granted directly to them, like parent role 1.

Role inheritance lets us define hierarchies between roles to simplify managing access grants. It also gives users the ability to utilize the privileges granted by multiple roles within a single transaction. Users can also individually assume roles indirectly granted to them through the hierarchy.

Why Should I Use Role Inheritance?

Role inheritance comes with many benefits. We can solve problems, like the one we identified in Chapter 7, where a user has read access to one dataset through one role and write access to a separate location using a separate and distinct role. Role inheritance simplifies access requests and grants, requiring very little work on the act of the database administrator. Using role inheritance also respects abstraction barriers and prevents us from repeating ourselves. We can use role inheritance to carefully curate particular types of roles and we can also use it to easily craft specialty roles.

Revisiting the problem we identified in Chapter 7, we can solve this using role inheritance. In the last chapter, we covered how we could solve that problem using secondary roles. Before we dive into a solution, let's revisit the problem. In Figure 9-3, we can see that user USER_1 has read access to schema SCHEMA_A through role ROLE_A, and write access to schema SCHEMA_B through role ROLE_B.

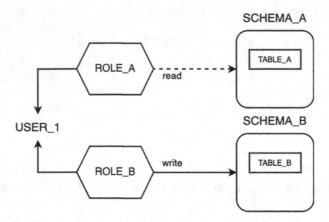

Figure 9-3. *User USER_1 has read access on schema SCHEMA_A and write access on schema SCHEMA_B*

When utilizing primary roles alone, user USER_1 cannot create a downstream table of table TABLE_A in schema SCHEMA_B because they cannot simultaneously use the privileges in role ROLE_A at the same time as using the privileges in role ROLE_B This is where role inheritance comes in. Instead of using role ROLE_A and role ROLE_B, we can grant role ROLE_A and ROLE_B to a new role ROLE_AB combining the privileges from both roles, as we can see in Figure 9-4.

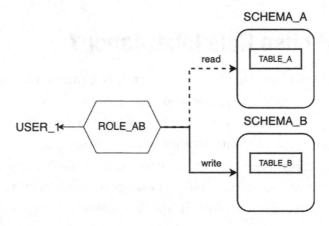

Figure 9-4. *Role ROLE_AB combines privileges from role ROLE_A and role ROLE_ B*

User USER_1 can now use role ROLE_AB as their primary role to successfully create a table in schema SCHEMA_B using data from schema SCHEMA_A. Similarly to secondary roles that we covered in the previous chapter, role inheritance can be used to glue all of our privileges together.

We can greatly reduce the workload of a database administrator (DBA) by using role inheritance. When we grant all of the access from one role to another, it only requires one line of SQL to grant and one line of SQL to revoke, if necessary. As we identified in the previous paragraph, a DBA would only need to know the parent role and the child role to manage the grants. The DBA would not need to be concerned with whether or not one single role would do everything a user needs – role inheritance creates that simplicity.

From a computer science perspective, I love role inheritance. We can create roles that respect abstraction barriers. Child roles – roles that are granted to other roles and that do not have roles granted to them, do not need to be aware of other child roles, the roles only govern their limited subset of data assets. Parent roles – those that are not granted any individual privileges, work as containers for low-level child roles. Since these roles have clear boundaries, this allows us to easily fit role inheritance into a programmatic way of managing roles.

We can use role inheritance to carefully curate particular types of specialty roles, which we will dive deeper into when we go through how to use role inheritance in the following section. If we want to create roles for a service account used by a particular tool, we can grant access to each dataset to a special role only used by the service account. This simplifies logging and ownership since the service account's role will appear as the owner of all objects it creates.

How Do I Use Role Inheritance?

When setting up role inheritance, there are a few concrete steps we need to take and considerations we need to make. Before we get into creating and distributing roles, we need to decide how we are going to create a hierarchy that works with the different access control frameworks. Once we decide on how we will set up the hierarchy, we can start composing roles. Roles can then be granted to other roles and to users. Finally, we will look at the end user experience and how users can benefit from role inheritance.

Creating a Hierarchy

Before we start creating and allocating roles, we need to come up with the right strategy for the organization. We will allocate roles differently depending on the paradigms we're using for creating roles. If we're using functional roles, team roles, or a combination of the two, then we will tailor our hierarchy accordingly.

Functional Role Hierarchy

When we create a hierarchy using purely functional roles, we need to take into account the access patterns that our users will have, and the different functional roles we have or anticipate having. In Chapter 5, we covered the different ways we can use functional roles. We can create functional roles for a dataset, and we can create functional roles for a particular job type or responsibility like engineering. In this situation, we want to identify atomic roles that are directly granted privileges on objects, and roles that are not granted privileges on objects, but are instead granted directly to users.

In the example of an engineer functional role and a few dataset functional roles, we can see how part of this hierarchy can be created in Figure 9-5.

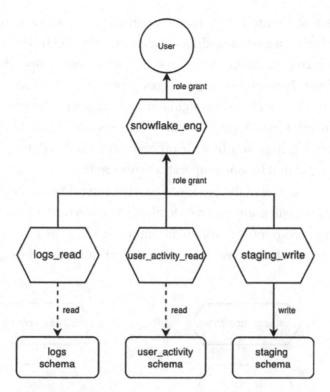

Figure 9-5. *A user is granted access to the snowflake_eng role which is granted three roles, giving the user access to three separate schemas*

In this diagram, we have three separate schemas, each with its own functional role. The logs schema has a read role called `logs_read`, the `user_activity` schema has a read role called `user_activity_read`, and the staging schema has a write role called `staging_write`. These three roles are atomic, meaning that they are low-level roles that are granted privileges directly on objects in the database. We also have the role `snowflake_eng`, which is granted all three of these roles but is not granted any direct object privileges. We can easily grant access to the three of these datasets to a user simply by granting a user the `snowflake_eng` role.

Team Role Hierarchy

Team roles combine with role inheritance very naturally. Since teams typically exist within a hierarchy, we can mimic that hierarchy using role inheritance. When we use role inheritance with team roles, we will actually *reverse* the existing hierarchy when we translate into Snowflake. To demonstrate this, we're going to work through an example using a revenue accounting team.

Let's say we have a revenue accounting team that reports into accounting, which is a part of finance, which is a part of business operations. We will have the roles revenue_accounting, accounting, finance, and business_operations, respectively. The revenue_accounting role should have privileges to objects only needed by the revenue accounting team, the accounting role might have privileges to objects that all of the accounting teams need, finance might include datasets used by all finance teams, and finally, business_operations may include other miscellaneous datasets occasionally used by all teams under that business operations umbrella.

Since we want to grant all of the business_operation role's privileges to everyone in that organization, we will grant that role to all of the roles under business operations in the organization's hierarchy. We will do the same with finance and accounting. In Figure 9-6, we can see the hierarchy of the company compared to the hierarchy of the grants.

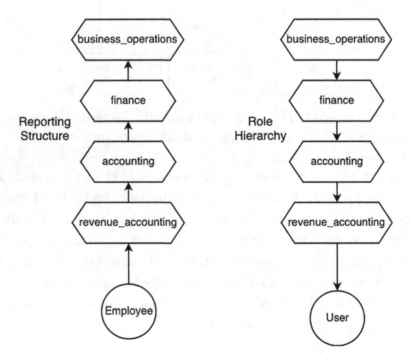

Figure 9-6. *The reporting structure starts at the employee and moves upwards, whereas the role hierarchy starts at the top and inherits downwards to the user*

As we can see in Figure 9-6, business_operations is granted to finance, which is in turn granted to accounting, which is then granted to revenue_accounting. The SQL to create this hierarchy would look like

```
GRANT ROLE BUSINESS_OPERATIONS TO ROLE FINANCE;
GRANT ROLE FINANCE TO ROLE ACCOUNTING;
GRANT ROLE ACCOUNTING TO ROLE REVENUE_ACCOUNTING;
```

Since these grants are recursive, a user with the revenue_accounting role would have access to the business_operations role, the finance role, and the accounting role, and access to the respective datasets. Since we execute grants on the way down from the highest-level role we have, we're not only executing the smallest number of grants, but we're also making it really easy to solve for a reorganization that may happen in the future.

Functional and Team Role Hierarchy

Organizations that utilize a combination of functional and team roles can benefit from using role inheritance to simplify the framework. The biggest benefit is if we have functional roles that every member of a team needs access to. In this situation, we can grant the functional roles directly to a team role. In Figure 9-7, we can see how we can set up a hierarchy using both functional and team roles.

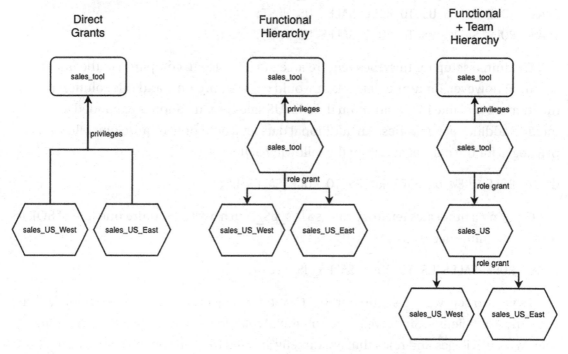

Figure 9-7. *Direct granting of privileges compared to creating a functional hierarchy, and a team and functional hierarchy*

In the first diagram, we can see how this would look if we directly granted access to the sales_tool schema to the sales_US_West role and the sales_US_East role. While this looks quite simple, each of those arrows represent between 5 and 20 lines of SQL, depending on the privileges granted to these teams. The second diagram shows how we could create a functional role for the sales_tool schema that is then granted to the sales_US_West and sales_US_East roles. This simplifies the grants because the privileges are only granted once to the sales_tool role, and then that role is granted to sales_US_West and sales_US_East. The SQL for this would be the privileges granted to sales tool and the following:

```
GRANT ROLE SALES_TOOL TO ROLE SALES_US_WEST;
GRANT ROLE SALES_TOOL TO ROLE SALES_US_EAST;
```

This solution requires roughly half the SQL required for the first solution.

The third diagram shows what this problem would look like if we had a hierarchy for team roles as well. Instead of granting sales_tool directly to sales_US_West and sales_US_East, sales_tool is granted to sales_US. The SQL for this initial grant would be

```
GRANT ROLE SALES_TOOL TO ROLE SALES_US;
GRANT ROLE SALES_US TO ROLE SALES_US_WEST;
GRANT ROLE SALES_US TO ROLE SALES_US_EAST;
```

The initial setup for this does require an extra line of SQL compared to the middle solution; however, for future datasets, it would only require one, and this solution is much more scalable in the situation that the US sales organization is expanded to include additional territories. An additional dataset would only require one role with the privileges for the dataset and then the following SQL:

```
GRANT ROLE <NEW_DATASET_ROLE> TO ROLE SALES_US;
```

Creating a new sales territory role, sales_US_South, would require one line of SQL to receive the same access as sales_US_West:

```
GRANT ROLE SALES_US TO ROLE SALES_US_SOUTH;
```

As we can see, we can set up our RBAC system using hierarchies to make our system scale more efficiently. Since we grant functional roles to team roles, we are separating roles with privileges and roles that are directly granted to users, which allows us to quickly grant and revoke access to data.

Another common use case for using functional roles and team roles is for analytics teams that serve stakeholders across multiple teams. In this case, instead of granting functional roles to team roles, we're going to grant team roles to functional roles. Let's work through this using the following example. We have a data analyst working on a centralized data team. This data analyst works with all data related to growth and marketing, two separate teams in our fictitious organization. All of our growth team's assets are accessible through a `growth_eng` role, and all of our marketing team's assets are accessible through the `marketing` role. We can create a functional role for this data analyst and others working on the same project called `data_analyst_acquisition`. We will grant both the `growth_eng` and `marketing` roles to `data_analyst_acquisition`. This way the analyst can see all of the data for both of these teams, without exposing the data across the two teams. We can visualize this in Figure 9-8.

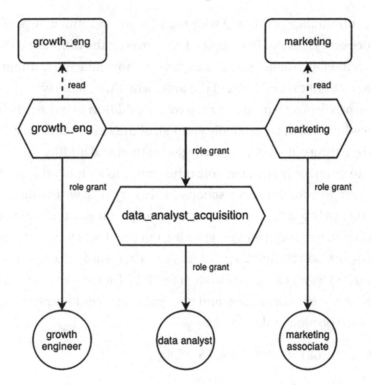

Figure 9-8. *The data_analyst_acquisition role combines the roles for the growth_ eng team and the marketing team*

In Figure 9-8, we can see how the `data_analyst_acquisition` role simplifies access for the data analyst at the same time as not cross-contaminating datasets between teams. The `growth_eng` role is directly granted access to the `growth_eng` schema, and the marketing role is directly granted access to the marketing schema, which are granted to users on their respective teams. Both the growth_eng role and the marketing role are granted to the data_analyst_acquisition role, which is granted to the data analyst.

   Team roles and role inheritance are very versatile, team roles can be combined in many different ways. The preceding examples are not exhaustive but are a demonstration of a few different ways these roles can be used with role hierarchies. One note of caution is that when creating these composite roles using inheritance, we may end up creating more roles than we need. If we are not careful of the different combinations we create, we may end up with many roles with only one user. To prevent this, I recommend querying the existing role grants before creating new composite roles.

Privilege Hierarchy

We can use role inheritance combined with read-, write-, admin-type privileged roles to simplify the process of creation. In Chapter 4, we covered the different types of privileges that exist in the Snowflake database. We saw that in Snowflake the number of privileges is higher than one might expect. Most of the time, when we create write access, we include all of the privileges granted to a read role in addition to write privileges, and when we create admin roles, we typically grant all of the read and write privileges as well. We can do this by listing out all of the privileges for read and adding them to our list of write privileges to grant, or we can use role inheritance to simplify this problem. Let's say we create a read role for the sales schema, sales_read, a write role, sales_write, and an admin role, sales_admin. If we have ten read privileges, ten write privileges, and five admin privileges, we would end up granting ten privileges to sales_read, granting 20 privileges to sales_write, and 25 privileges to sales_admin, for a total of 55 grants. If we use role inheritance, we can get that down to 27, 25 for the privileges granted directly to roles – a number we cannot reduce, and two grants of roles to other roles. The SQL for the role inheritance portion looks like

```
GRANT <READ PRIVILEGE> TO ROLE SALES_READ;
...
GRANT <WRITE PRIVILEGE> TO ROLE SALES_WRITE;
...
GRANT <ADMIN PRIVILEGE> TO ROLE SALES_ADMIN;
...
GRANT ROLE SALES_READ TO ROLE SALES_WRITE;
GRANT ROLE SALES_WRITE TO ROLE SALES_ADMIN;
```

We can also visualize the relationship between these roles in Figure 9-9.

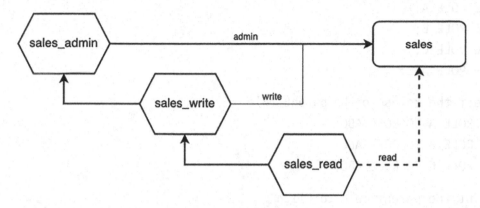

Figure 9-9. *The read role and privileges are granted to the write role in addition to write privileges, which are then granted to the admin role and combined with admin privileges*

As we can see, we greatly reduce the number of grants when using role inheritance to create a hierarchy of privileges.

Role Inheritance and Specialty Roles

We can use role inheritance to create specialty roles. In an earlier section, we explained how we could easily create a role for our ETL or ELT process's service account. We can create a role like etl_airflow, which can be granted the dataset_write role for every respective dataset to be managed by this process. This allows us to only grant access to necessary objects and limit the scope of the service account. We can repeat this process for reader accounts, accounts we create for third party tools, and any other roles that require a less-than-standard set of privileges across the database.

End User Experience

Since we care about the end user experience as well as the developer experience, we need to know how end users interact with inherited roles. We're going to use an example with a new user account for my best friend Ellen, who isn't a technical user. We're going to create a few roles, grant them to a parent role, grant the parent role to Ellen, and then check what that looks like at her end.

```
// create the roles
CREATE ROLE A;
CREATE ROLE B;
CREATE ROLE C;
CREATE ROLE ABC;

// grant the roles to the parent role
GRANT ROLE A TO ROLE ABC;
GRANT ROLE B TO ROLE ABC;
GRANT ROLE C TO ROLE ABC;

// grant the parent role to Ellen
GRANT ROLE ABC TO USER ELLEN;
```

When we login to the user interface using Ellen's account, we see the screenshot in Figure 9-10.

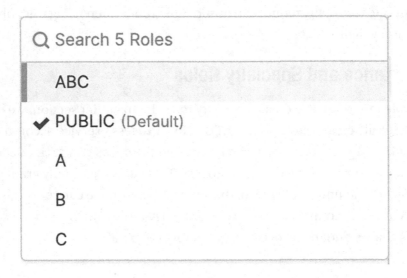

Figure 9-10. *The drop-down selector for roles for Ellen*

She can see all four of the roles she has access to, in addition to the default role granted to all users, PUBLIC. This is when naming conventions become very important – we should name the parent roles in such a way that it is clear to the end user which one to use since all child roles will also be available to them.

Naming Conventions

When we use role inheritance to create roles, we will necessarily create a multitude of roles, and at the same time, most users will have many roles at their disposal as well. As we covered in the previous section, end users will need to know which roles to use, and if we name our roles properly, the name alone should indicate that. The easiest option is to name our child roles that are directly granted privileges with names that indicate that they should not be used, and to name our composite or parent roles a name that corresponds with job functions or team names.

For child roles directly granted privileges, I recommend using a naming convention that indicates the types of privileges the role has. We've seen this throughout the book as well as in roles such as SALES_TOOL_READ. We could make this role even clearer by naming it something like DIRECT_SALES_TOOL_READ. Ideally, the corresponding role that a user should assume would have a name like SALES_ANALYST or SALES_TEAM. The words DIRECT and READ make the name of the role harder for less technical users to immediately decipher, which makes them less likely to reach for that role when roles with simpler, more relevant names are available that may match their team name or their job title.

Role Inheritance and Secondary Roles

Role inheritance and secondary roles can be used together; however, their benefits do overlap somewhat and as such this limits the benefits of combining them. It may behoove your organization to use role inheritance without using secondary roles, whether on a tool-by-tool basis or completely. One benefit to using role inheritance without secondary roles or using role inheritance in an integration or system that does not have secondary roles enabled is that users can assume the inherited roles separately, as well as combined. In a situation where a data analyst creates a dashboard that will be viewed by users with multiple different roles, the data analyst can visualize the results for each of the different roles they're granted.

Using the Snowflake UI, an analyst can easily toggle the different roles assumed using the dropdown role menu, as shown in the preceding section, or through using the USE ROLE command. An analyst creating views or downstream tables can verify that their stakeholders have appropriate access by assuming each of those roles and validating that the query returns the correct data for each of these user groups.

As we covered in the last chapter, secondary roles can be enabled on an integration-by-integration basis, and on a user-by-user basis. This means that we can exempt certain tools either by not enabling secondary roles, or some tools can be exempt because they do not yet support secondary roles. As we covered in Chapter 7, some tools allow users to specify a primary role, and some rely on default roles, and so those may not allow a user to assume the roles their stakeholders may use to interact with data. In the situation I described in the last chapter, where we enabled secondary roles to fix authentication through Tableau, we chose to enable secondary roles on Tableau Online, but leave them disabled on Tableau Desktop so that analysts could check access before deploying.

Logging with Role Inheritance

With role inheritance, just like with secondary roles, since we are essentially skirting around the boundaries placed by using a primary role, logging might look differently than one would typically expect. Let's run through a quick example using a child role to create a schema, then switch to the parent role to create a table in that schema.

```
// creating the role
CREATE ROLE CHILD_ROLE;
// granting it to myself
GRANT ROLE CHILD_ROLE TO USER JESSLARSON;
// make this my primary role
USE ROLE CHILD_ROLE;
// create a schema owned by child_role
CREATE SCHEMA CHILD_SCHEMA;
```

At this point, we now have a role CHILD_ROLE, which has been granted to myself and a CHILD_SCHEMA owned by the CHILD_ROLE. Next, we're going to grant CHILD_ROLE to a parent role, and create a table using that role.

```
// switch back to accountadmin for creating a role
USE ROLE ACCOUNTADMIN;
// create parent_role
CREATE ROLE PARENT_ROLE;
// role inheritance! Parent role inherits child role
GRANT ROLE CHILD_ROLE TO ROLE PARENT_ROLE;
```

```
// grant role to myself
GRANT ROLE PARENT_ROLE TO USER JESSLARSON;
// assume parent_role as primary role
USE ROLE PARENT_ROLE;
// create a table
CREATE TABLE CHILD_SCHEMA.TEST AS (SELECT 1 AS NUM);
```

Now we have PARENT_ROLE inheriting CHILD_ROLE and all of its privileges, which I have granted to myself, and used to create a table TEST. Next, we're going to query for the grants to this table to see who owns it.

```
SHOW GRANTS ON TABLE CHILD_SCHEMA.TEST;
```

This gives us the output in the screenshot in Figure 9-11.

created_on	privilege	granted_on	name	granted_to	grantee_name	grant_option	granted_by
2021-09-23 ...	OWNERSHIP	TABLE	PROD.CHILD_SCHEMA.TEST	ROLE	PARENT_ROLE	true	PARENT_ROLE

Figure 9-11. *Output from querying the grants on the child_schema.test table*

This shows us that the owner of the table is the primary role we used, PARENT_ROLE. This makes sense because the primary role is the default owner of new objects. We can compare the output from the table's grants to the output from the grants on the schema.

```
SHOW GRANTS ON SCHEMA CHILD_SCHEMA;
```

This gives us the output in the screenshot in Figure 9-12.

created_on	privilege	granted_on	name	granted_to	grantee_name	grant_option	granted_by
2021-09-23 ...	OWNERSHIP	SCHEMA	PROD.CHILD_SCHEMA	ROLE	CHILD_ROLE	true	CHILD_ROLE

Figure 9-12. *Output from querying the grants on child_schema*

As we can see, the only role that is granted privileges on this schema is CHILD_ROLE. This conflicts with the grants on the table CHILD_SCHEMA.TEST. It is important to keep this in mind as it can create discrepancies between the explicit grants on objects, and the ownership of objects inside those containers.

Key Takeaways

Role inheritance is a very powerful tool we can use to manage our role-based access control in a way that is scalable. It allows us to easily adjust for organizational team changes as well as changes in responsibility. Additional key takeaways from this chapter are listed here:

- We can use role inheritance to grant all of the privileges from a role to another role, in addition to the role itself.

- Role inheritance allows users to stitch together privileges from different roles during a single transaction.

- Role hierarchies can be created differently for functional roles, team roles, and a combination of both.

- Analysts can use role inheritance to verify the output that stakeholders would see before publishing a dashboard.

- Creation of objects while using role inheritance creates discrepancies between grants on that new object and the container.

In the next chapter, we will shift gears and start to cover the privileges that exist within the different stages in the object hierarchy.

CHAPTER 10

Account- and Database-Level Privileges

We've covered the different privileges that exist in Snowflake, how to think about organizing access, and how to curate roles so that they best work with your organization. Now we're going to focus on actually allocating privileges to roles in a systematic way at all levels within the database hierarchy. We can allocate access to objects at the highest level – the account level – and we can allocate access at the underlying levels – database, schema, table, row, and column – which we can visualize in Figure 10-1.

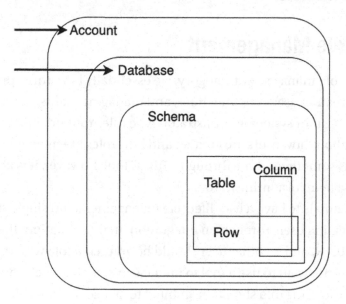

Figure 10-1. *The different levels at which we can allocate access in Snowflake*

© Jessica Megan Larson 2022
J. M. Larson, *Snowflake Access Control*, https://doi.org/10.1007/978-1-4842-8038-6_10

We're going to start with access at the account level, and move our way into the smaller, more granular access types. This way, we can make sure that the decisions we make at the beginning are consistent across the board. Since objects located at the account or database level are not atomic type objects that hold data, most of the privileges, and as a result, roles, allocated privileges at this level, will be administrative or specialized roles for monitoring and logging.

Account Level

We're going to start at the account level, which we've referred to as the global level in previous chapters. The account-level privileges are less oriented around data and more oriented about account-level attributes like user management and billing. However, revisiting Chapter 3, this becomes important for things like the United States' Sarbanes-Oxley internal controls. We can break down account-level privileges into four distinct categories: user and role management, creating account-level objects, monitoring activity, and miscellaneous.

User and Role Management

For the user and role management category, we essentially have three privileges we need to think about: creating users, creating roles, and managing grants. These all typically belong to an account- or system administrator–type role. Your organization will need to make decisions about how users are created and how roles are managed. Creating and managing roles is sometimes done through a RBAC tool; however, it is also common to see this done by a team of administrators.

A common setup is to have a user lifecycle management and single sign-on tool own the process of creating users through an integration. In that situation, the tool would be granted CREATE USER, and no other user should be granted a role with this access. If your organization does not want to use a tool to manage this, then a role should be created with this privilege and the role should be granted to all users that will be creating users.

Whether your organization uses a tool to manage access grants or whether your organization has a team of administrators that manage roles, a role with the CREATE ROLE and MANAGE GRANTS privileges should be created. If the same group of people creates users that manages grants, all three of these privileges should be granted to that same role.

Creating Account-Level Objects

We can separate out the privileges for creating account-level objects from the other account-level privileges. The following privileges grant the ability to create net-new objects at the account level:

- CREATE WAREHOUSE
- CREATE DATA EXCHANGE LISTING
- CREATE DATABASE
- CREATE INTEGRATION
- CREATE NETWORK POLICY
- CREATE SHARE
- IMPORT SHARE
- OVERRIDE SHARE RESTRICTIONS

With the exception of CREATE DATABASE and CREATE WAREHOUSE, all of these involve security-related decisions. I would separate privileges into two separate roles; CREATE DATABASE and CREATE WAREHOUSE should be privileges granted to an administrator-type role, someone who would make decisions about how the database is structured, and how warehouses are distributed among teams and tools. The rest of these privileges should be granted to a security role. In this situation, a centralized data platform team may make the decision to create a new network policy, which would then go to a security team for approval and the actual creation of the network policy.

Monitoring Activity

We can separate out the monitoring privileges from the other account-level permissions and allocate these to a logging or metrics role. The privileges associated with this are MONITOR EXECUTION and MONITOR USAGE. A sensor that checks the status of an asynchronous task like loading data from a pipe, might need the MONITOR EXECUTION privilege. I recommend granting both of these privileges to a monitoring or logging role.

Miscellaneous

The fourth category of account-level metrics is kind of a catch-all bucket. These are essentially the privileges associated with actioning certain types of objects. The privileges are the following:

- APPLY MASKING POLICY

- APPLY ROW ACCESS POLICY

- APPLY TAG

- ATTACH POLICY

- EXECUTE TASK

These privileges may be necessary for a tool that does RBAC, if your organization uses a tool to handle that. Otherwise, these privileges might make the most sense for a database administrator managing grants. Applying row access policies and masking policies might be something non-admin users will need if they're managing fine-grained access on their own.

Database Level

There are four main privileges allocated at the database level, which we can split into three role categories. We can segment these privileges into the following categories: read, admin, and monitor. For read access, we can grant USAGE on that database to a group of users who will not be allowed to create schemas or modify the database but will need to use objects within the database. For admin access, we can grant MODIFY and CREATE SCHEMA, allowing administrative users to make changes to the database itself and to create schemas. Finally, we can allocate MONITOR to a logging role, but also to roles that do not have read access on the database, so that they can see the names of the schemas located within the database, without having access to any of the underlying data.

It's important to note that everyone who will use the database will need to be granted USAGE. Without USAGE on a database, users will not be able to interact with any of the underlying schemas, tables, or views.

Key Takeaways

At the account and database level, we can manage objects that exist at the highest levels of the object hierarchy. Many of these account-level objects, such as warehouses, affect objects much lower in the hierarchy. In the upcoming chapters, we will dive deeper into the objects that exist in Snowflake, starting with schema-level privileges in the next chapter. The following are key takeaways from this chapter:

- Account-level permissions and database-level permissions are primarily for administrative and specialized roles.

- Monitoring and logging privileges primarily exist at the account and database level.

- All users who need to use a database will need usage on that database to use any of the underlying schemas.

CHAPTER 11

Schema-Level Privileges

In the last chapter, we covered access control at the account and database levels. Since no proper data exists at those levels, the access control decisions are more geared towards administrative roles, specialized roles like monitoring and logging, and setting up access for downstream data. In this chapter, we're going to cover permissioning at the schema level, as shown in Figure 11-1.

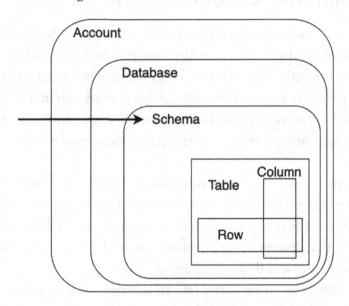

Figure 11-1. *Generally, the levels of access that exist in Snowflake*

The schema level is the smallest container we have for first-order data like tables and views. As a result, most of the access control decisions that we need to make at the schema level will be around the objects contained in the schema.

© Jessica Megan Larson 2022
J. M. Larson, *Snowflake Access Control*, https://doi.org/10.1007/978-1-4842-8038-6_11

What Is Schema-Level Access?

A schema is the smallest container we have for tables and views. When we allocate access at the schema level, we are allocating access to groups of objects at a time, rather than individually and directly granting access to objects. Since everything is granted on the schema, and privileges are granted to one role on one object type at a time, every object of the same type will have the same privilege for the same role. We can have different privileges for views and tables for a given role in a schema, but we cannot have different privileges on different views for the same role and still call it schema-level access.

Why Use Schema-Level Access?

Schema-level access is a common pattern because it fits well with existing systems. It is common for users to need access to an entire dataset, which is usually stored as a series of tables and views within a single schema. We can mirror our team and functional roles with the schemas we create to simplify administration. Schema-level access scales efficiently compared to more granular grants like table level, requiring fewer overall grants. Most organizations can benefit from using schema-level access to manage some of their data assets.

Since users typically work with an entire dataset, rather than pieces of it, using schemas can be the easiest and most logical solution. One dataset is typically stored in one or more schemas, and users are granted access to all of the tables in that dataset at a time. This means that instead of granting access to each table individually, the entire schema or set of schemas and dataset is granted at one time.

We can mirror our roles so we have a one-to-one mapping between roles and schemas. This makes managing schema-level access easy to do programmatically because each dataset has a role and schema with the same name. This also makes it easy to manage without the use of scripts because it takes away the guesswork. If there's a sales tool schema, `sales_tool`, and roles `sales_tool_read` and `sales_tool_write`, a user simply asks for access to the sales tool dataset, and the database administrator can easily identify the schema and role to assign to that user.

Schema-level access scales really well because it requires one order of magnitude fewer grants than if done at the table level. Instead of granting access to each table individually, groups of tables and views are granted at one time. We can also take advantage of bulk grants like granting a privilege on all future tables or granting a privilege on all existing tables.

Schema-Level Privileges

We can segment the different schema-level privileges into a few categories similarly to how we split up the different account and database privileges. I like to split these up by function, based on what type of a user would need these privileges. This leads me to the following categories: administrative, monitoring, read, write, and data engineering platform.

For more detail about what these privileges mean, I recommend revisiting Chapter 5. This chapter covers these privileges in more depth.

Administrative Privileges

Schema-level privileges can be put into buckets for roles in a few different ways. For the administrative privileges, it makes the most sense to me to grant an admin the ability to MODIFY a schema. I would also grant the schema administrator the OWNERSHIP privilege. Since these privileges are related to the container and not the use, creation, or deletion of any data, these make the most sense for an administrative role rather than a write role.

Managed Access

Managed access schemas differ from regular non-managed schemas in that the owner of a managed access schema does not automatically have privileges to grant access on the schema or any of its objects to other roles. Roles such as SECURITYADMIN will retain the ability to manage grants on these schemas, as well as roles with the MANAGE GRANTS privilege. To create a managed schema, we simply add the option to the creation of a schema, as we can see here:

```
CREATE SCHEMA <SCHEMA NAME> WITH MANAGED ACCESS;
```

This option is recommended for organizations that would like to delegate all granting of privileges to specific administrator roles rather than allowing administrators on a dataset level.

Monitoring Privileges

As with the database- and account-level permissioning, schema-level permission carves out privileges for a monitoring or logging role. I would grant the same monitor or logging role from higher up in the hierarchy, or a monitoring role with a more limited scope the MONITOR privilege, in addition to the USAGE privilege. Another use-case for this is to grant monitor to all users in the organization on all schemas, so that users can see metadata associated with the data that exists in Snowflake, without exposing the actual data. Typically, an organization might handle this with a data governance or data cataloging tool; however, granting monitor on all schemas to all users doesn't cost anything and provides a limited amount of visibility without much work.

When granting the MONITOR privilege, it is important to note that since a schema is just a container for objects, rather than an object itself, meaning that just granting monitor will not allow a user to see all objects within a schema. Monitor simply allows the operation DESCRIBE SCHEMA <SCHEMA_NAME> to be run successfully. This allows a user to see all of the tables and views in a schema, which may be useful to identify needed datasets. For a table to appear in the results of that query, the REFERENCES privilege must also be granted to a role. We will cover this more in the bulk grants section later in this chapter.

Read Privileges

Read privileges do not differ significantly from monitoring privileges. A read role will need USAGE on a schema to be able to interact with any underlying tables or views. As mentioned earlier, a role will also need privileges on any schema object it will use.

Write Privileges

Granting write privileges at the schema level is a bit more straightforward than the other privileges granted at the schema level. In addition to the read privileges from the last section, the following privileges should be granted to a write role for the schema:

- CREATE TABLE
- CREATE EXTERNAL TABLE
- CREATE VIEW

- CREATE MATERIALIZED VIEW

- CREATE MASKING POLICY*

- CREATE ROW ACCESS POLICY*

I associate these privileges with a standard write role because they're the standard objects that are most commonly used. There isn't anything particularly tricky or unique about tables and views, compared with some of the other objects in Snowflake. It makes sense for a role writing data to a schema to need to create tables or views with some frequency.

Additionally, I might include the ability to create masking or row access policies in the schema write role, though they may not necessarily belong in a true write role. These are typically more administrative privileges, and may make more sense for a data engineering platform role or an administrator role, depending on your organization's preference. The user or group of users creating tables or views in a schema will likely be the subject matter experts, or close enough to subject matter experts, that they will be the most well equipped to make decisions about how fine-grained access control should be implemented on a table-by-table basis. Remember that the privilege for applying these policies once they're created is granted at the account level, as discussed in the last chapter. A grant to an analyst managing row access policies would look like

```
GRANT APPLY ROW ACCESS POLICY TO ROLE <ROLE NAME>;
GRANT CREATE ROW ACCESS POLICY ON SCHEMA <SCHEMA NAME> TO ROLE <ROLE NAME>;
```

Both of these grants, in addition to USAGE on the database and schema, should be sufficient for a user to create a row access policy in that schema, so long as the user has the correct access to objects referenced within the policy definition or mapping table.

Data Engineering Platform Privileges

This category of privileges is essentially everything that doesn't neatly fit into the preceding categories. This is a group of objects and tasks that can be standardized and created by a centralized team. These privileges include

- CREATE STAGE

- CREATE FILE FORMAT

- CREATE SEQUENCE

- CREATE FUNCTION

- CREATE PIPE

- CREATE STREAM

- CREATE TASK

- CREATE PROCEDURE

- CREATE MASKING POLICY*

- CREATE ROW ACCESS POLICY*

Creating a stage requires knowledge of the integrated storage locations, which is typically something that lies on the data engineering or security side of the house. File formats also require some knowledge about how the raw data is stored, again, more of a data engineering task. Sequences require more in-depth knowledge around how tables relate to each other. Functions can be defined in SQL, but also in Java and JavaScript, making this more of an engineering task. Pipes, like stages, require some understanding of the storage, and may require setting up messaging service infrastructure. Streams do not hold any underlying table data, so it is unlikely that they would be useful for an analyst user.

All of the preceding objects require a little more knowledge typically associated with a data engineer or security engineer, which makes them good candidates for standardized objects provided to users outside a centralized data team. If a team needs another file format, for example, it might make the most sense for them to ask the centralized data team to create that file format. This provides a more consistent experience for everyone and prevents a free-for-all with many duplicate objects.

Additionally, creating masking policies or row access policies can be granted to a data engineering platform role rather than the write role. These privileges fit somewhere between platform administration and a true write role, and as such, the role with these privileges may change depending on how a particular organization uses Snowflake. If your organization does not want analysts to manage these, then it may make more sense when these privileges are granted to a data engineering platform role

Bulk Grants on Schema Objects

There are two main ways to handle privileges on schema objects. The first way is to have the main role own a schema and create all objects within it. The second way is to use bulk grants to manage many objects at a time. There is no right or wrong answer to which of these works best, as with everything else with database design, it depends on the organization.

When using the ownership way of managing privileges on schema objects, the same role is used to create all objects within a schema. Since they're all created using the same role, that role automatically has all privileges on each of these objects. This works great if there is a one-to-one mapping between roles and schemas. This does not work well if there is more than one role used on a given schema. Since the same role must be used to create an object as reading from an object, it is impossible to use this method and also utilize read-only roles. As a result, it often makes sense to utilize bulk grants.

Bulk grants allow privileges to be granted on multiple objects at a time, as well as on objects that do not yet exist. This greatly simplifies access to objects and allows us to maintain schema-level access. There are two main types of bulk grants: grants using the all keyword, and future grants.

All

We can grant access to all existing objects within a schema. This works great when retroactively cleaning up access control where a schema already has tables and views, and when creating a new role to apply to an existing schema. To grant access to all objects in a schema to a role, we can run a query like the following:

```
GRANT <PRIVILEGE> ON ALL <OBJECT TYPE PLURAL> IN SCHEMA <SCHEMA NAME> TO
ROLE <ROLE NAME>;
```

Since privileges are specific to the type of object, we can grant a privilege on all of one type of object to a role at a time. This only affects objects that currently exist, so if future objects will exist, it is best to pair the all bulk grants with a matching set of future grants.

Future Grants

Future grants allow us to grant a privilege on all future instances of a type of object within a schema to a role. It is important to note that this only affects objects created within this schema and will not affect objects created in a different schema and later moved to this schema. We can grant a privilege to all future objects of a certain type to a role by running the following SQL query:

```
GRANT <PRIVILEGE> ON FUTURE <OBJECT TYPE PLURAL> IN SCHEMA <SCHEMA NAME> TO
ROLE <ROLE NAME>;
```

Just like when we use bulk grants with the all keyword, we should pair future grants with grants to all existing objects if there are any existing objects. For this reason, I always grant privileges using both methods. This means that by default there will be consistent privileges between all objects of the same type within a schema.

Mapping Roles to Schemas

When we create roles for access to specific schemas, it is important to use a consistent naming convention, as well as to choose the ideal mapping patterns for the use cases at our organization. If we're using team roles, it often makes sense to have a team schema that serves as a sandbox for users. If we're using functional roles, we might want to create multiple schemas for the roles so that there is a delineation between different states of data.

Team Roles

When we use a team role framework for access control, we should think about having a team sandbox schema. By creating a sandbox schema for each team, we give users somewhere to put tables and views that may not be designed for a wider audience. One example of this is for development. A team schema can be used as a way to create pseudo development and production separation. Since these assets are in a team's schema and not visible to a wider audience, communicating that an asset is a work in progress is an easier task. An organization might also decide on creating multiple schemas for each team, so that each team can have a schema for completed and supported objects and a schema for miscellaneous projects in progress. In any situation, having a team schema where teams can feel free to create whatever they'd like helps increase development velocity.

Functional Roles

When creating schemas for functional roles, many of the same concepts from team roles hold. While we may not want to have sandboxes for functional roles since they can be used by members from multiple teams, exposing miscommunication risks, we can separate data that comes in raw, straight from the source, from data that is created downstream of the raw data. If we have a dataset from sales tool, instead of creating one `sales_tool` schema, we can create two schemas: `sales_tool_raw` and `sales_tool_denorm` for denormalized tables. We can visualize this in Figure 11-2.

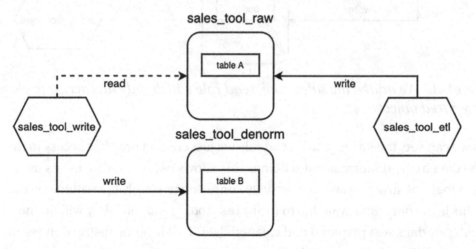

Figure 11-2. *The sales_tool_write role has read access on the raw schema and write access on the denormalized schema. The sales_tool_etl role used by the pipeline has write access on the raw schema, and no access to the denormalized schema*

As we can see in Figure 11-2, two different `sales_tool` roles exist. The write role used by analysts to create downstream assets has read-only access on the raw data schema and has write access on the denormalized schema. This allows the users to pull data from the raw schema and to create downstream assets in the denormalized schema. The `sales_tool_etl` role is only granted to the machines or tools that deposit data into

Snowflake from an external source. This role has no access to the denormalized schema as it does not have involvement with downstream assets. This allows us to keep a clear separation between raw data and cleaned assets.

Additionally, we could have a read role used to create dashboards in a visualization tool that only has access to the cleaned data schema. We can visualize this in Figure 11-3.

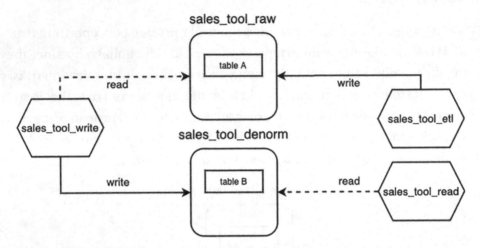

Figure 11-3. *We added the sales_tool_read role which only has access to cleaned and prepared objects*

As we can see, the `sales_tool_read` role we just created only has access to the cleaned data in the denormalized schema. This allows users to create assets using business logic, abstracting away potentially confusing or poorly formatted raw data. Since this is the only data available to the `sales_tool_read` role, they will not need to deal with raw data, just prepared and cleaned data. This layer of abstraction essentially creates an Application Programming Interface (API), the cleaned data schema abstracts away the unnecessary details and complexity that may be present in the raw data.

Specialized Schemas

Since schemas are simply containers for different objects of all types, and objects can be used across schemas, we may have a need to create specialized schemas for different functions. Depending on how we set these up, we may need to restrict schema ownership to a centralized administrator role. We can create resource schemas to store commonly used objects like file formats or stages. Security schemas can be used to store access policies, which might only be standardized by a security team. Another specialized schema could hold logs generated by random processes within Snowflake.

If we want to restrict all of a type of object to a certain schema, we must lock down all creation of that type of object outside of that schema. Since OWNERSHIP on a schema grants all privileges, we will need to restrict ownership of schemas to a platform administrator role. When we do this, we grant all privileges except the ability to create this type of object to the "owning" role. If we don't do this, then we will not be able to restrict all objects of a specific type to a particular resource schema.

By creating resource schemas for certain types of widely used objects, we can standardize the objects and have a centralized team responsible for all of these objects. An example of this would be to create a schema for file formats, file_formats. In this situation, we could have a data_engineering role with admin access on file_formats tasked with creating standardized file format objects. Any other role that will be using these file formats should get usage on all of these assets in the schema.

In much the same way as these resource schemas, we could have one or more security schemas used to store objects related to security and data privacy. We might have a schema access_policies that stores all row access policies and all column masking policies. The security team should administer this schema and create all of the policy objects. Any team that might need to apply these policies to tables and views will simply need usage on these policies, and the global apply privileges.

We might also create logging schemas for scripts or tools to use. One example might be for an ETL process that automatically evolves tables to match the input structure if it changes. Whenever this process changes the structure of the table, it logs a timestamp and the DDL (Data Definition Language) used to make the changes to the table. Read privileges can be granted to a logging service account log to an external logging service as well.

Key Takeaways

Managing privileges at the schema level allows us to easily set the same privileges on a dataset-by-dataset basis. If we wish to set privileges at a lower level than the schema, we must still allocate privileges on the schema level, as without it, users will have no privileges on the schema objects. In the next chapter, we will focus on table- and view-level privileges. The following are key takeaways from this chapter:

- Schema-level access allows us to manage privileges on many objects at a time.

- Since datasets are often stored as multiple tables in a single schema, schema-level access can easily fit a business's practices.

- Granting access to objects on the schema level simplifies administration over granting at the object level.

- Bulk Grants can be used to grant privileges on all existing or future objects in a schema.

- Specialized schemas can be created to store certain object types, creating a centralized location for them.

- The ownership privilege on a schema can be tricky, make sure that is the right fit before granting it.

- We can create managed schemas that remove the ability to grant additional permissions on schema objects from the owning role.

CHAPTER 12

Table- and View-Level Privileges

In the previous two chapters, we covered implementing access controls at the account, database, and schema levels. Privileges granted at these levels affect many objects at a time, since an account, database, and schema are containers for other types of objects. In this chapter, we're going to cover implementing access controls at the table level, as shown in Figure 12-1.

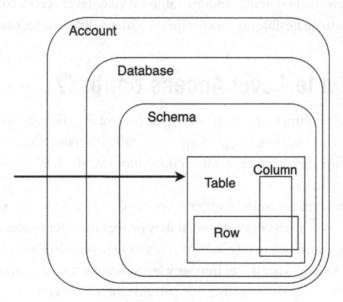

Figure 12-1. *The levels of access that exist in Snowflake*

As you can see, this is not the lowest level of access that we will work with in Snowflake. However, table level is the first level we will work with that directly holds data. In this chapter, we will cover what table-level access is, why an organization may choose to grant privileges at this level, and how to implement it.

© Jessica Megan Larson 2022
J. M. Larson, *Snowflake Access Control*, https://doi.org/10.1007/978-1-4842-8038-6_12

What Is Table-Level Access Control?

Table-level access control involves granting privileges on the table or view level rather than at the schema level. This means that we can grant privileges with greater precision, but at the same time, with that precision comes a lot more management. We can also adopt a combination of schema-level access and table-level access.

Managing table-level access requires more administrative work than schema-level access; the cardinality of the number of tables can be orders of magnitude larger than the number of schemas. We can use a few different strategies to simplify table-level access. We can reduce the overall number of schemas, meaning that the schema-level grants will not need to be repeated in addition to the table-level grants. Another strategy is adopting a hybrid schema and table-level permissioning system.

By combining schema-level and table-level permissioning, we can simplify access requests, while at the same time utilizing table-level permissioning when we require that granularity. We may decide that the majority of our data can be managed at the schema level, but that a few select schemas require table- or view-level access controls. This strategy allows us to be flexible and accommodate many different access patterns.

Why Use Table-Level Access Control?

Organizations that require more stringent access control may benefit from table- and view-level access control. Granting privileges on a table-by-table basis allows greater precision. There are also certain special circumstances in which table-level access control can be very useful.

Certain use cases make table-level access control a better fit for an organization than utilizing schema-level access everywhere. It may be that the data model organizes tables in such a way that teams will need a subset of tables across multiple schemas. It may be that your organization groups tables into very few schemas, and access control is being applied retroactively. It is also possible that certain tables hold very sensitive data, and for those tables, column- or row-level access is not ideal. In any situation, there may be reasons that access control needs to be administered on a table-by-table basis.

In addition to organization-wide requirements for table-level access, organizations may want to handle certain subsets of tables with table-level access. One special circumstance might be a workflow that allows a third party to access Snowflake data using a service account. This service account might only need data from one table; in this situation, granting very limited access to this single table is the safest, most secure way to handle this. Another potential use case is for assets managed by a centralized data team, or a cross-functional team of analysts providing data assets for many different stakeholder teams. This team or teams might have a schema where they create all of these assets, and then they can grant read access on a table-by-table basis to the respective stakeholder team.

Whether an organization implements table-level access control at the global level, spanning all data assets, or they implement it in a piecewise format, many organizations can solve problems using table-level access.

What Are the Different Types of Views?

You may have noticed that there are three different types of views in Snowflake: traditional views, materialized views, and secure views. The three different types of views behave differently from an access control perspective and have different implications.

Standard Views

Standard views are essentially stored SQL that behaves similarly to a table for an end user. Anyone with the REFERENCES privilege on a view can see the SQL used to generate the view, including the structure. The results fetched by a view will reflect the privileges of the owning role.

Materialized Views

Materialized views appear to behave the same way to an end user. It is only when we look under the hood and investigate further that we can see the differences between a materialized view and a standard view. Querying a materialized view may be more performant than a standard view, because the data is pre-computed and stored in Snowflake. Since the data is pre-computed, it may not always be up to date, meaning that any updates to logic that controls granular access control may not go into effect immediately. Like with standard views, the privileges of the owning role will dictate the data returned.

Secure Views

Secure views are best suited for use with granular access control. This is because secure views do not expose underlying data during internal optimizations, and do not allow users to view the definition SQL unless they're granted the OWNERSHIP privilege.

Table- and View-Level Privileges

We briefly went over the different table- and view-level privileges in the last chapter when we touched on bulk schema grants. We also went into this in greater detail in Chapter 4 when we covered all the privileges that exist in Snowflake. Tables and views share some of the same privileges, though some differ, as the nature of the two objects is inherently different.

Since we went in depth on privileges in Chapter 4, I'm going to cover the privileges with less depth in this chapter. We can categorize the table- and view-level privileges into a few categories: read, write, and administrator. It is also important to remember that a user will need usage on the database and schema that each table lives in, in addition to the privileges granted here.

Read Privileges

To grant read access to a table or view in Snowflake, we can grant a subset of the privileges. For a table or view, including external tables, materialized views, and secure views, a read role should be granted SELECT and REFERENCES. This grants the ability to query data from a table and view the table metadata using the DESCRIBE command. We can grant these privileges on a table like so:

```
GRANT SELECT, REFERENCES ON TABLE <FULLY QUALIFIED TABLE NAME> TO ROLE
<ROLE NAME>;
```

As we can see, the syntax for granting privileges on a table or view is similar to how we grant privileges at the schema level.

Write Privileges

Granting write access on a table or view requires the ability to update or modify a table. Remember that the ability to create a table or view is granted at the schema level, not at the object level because you cannot grant access on an object that does not exist without using the future bulk grant. A write role should have the following privileges on a table or view: INSERT, UPDATE, DELETE, and TRUNCATE, in addition to the read privileges. Some organizations might decide that a write role should be able to add data but not delete data, or that it can add data and modify a table, but not drop a table. In those situations, it might be better to grant INSERT and UPDATE to the write role and DELETE and TRUNCATE to the admin role.

Admin Privileges

The administrator of a table or view should have OWNERSHIP. This is required for most ALTER commands, and to drop a table or view. In addition to ownership, the admin should be granted the ALL privilege, meaning that they are able to do all operations that exist on an object. It is important to note that only one role can have ownership of a table, so if other roles need the ability to modify certain characteristics of a table, or need to drop and recreate a table, then a new role will need to be created for that group of users.

How Do I Implement Table-Level Access Control?

Granting privileges on a table or view is relatively simple. Since each of these grants will only affect one object, it is unlikely for unexpected ramifications to occur. To grant a privilege on a table, run the following command:

```
GRANT <PRIVILEGE> ON TABLE <TABLE NAME> TO ROLE <ROLE NAME>;
```

Privileges can also be chained together to simplify the number of grants executed like

```
GRANT <PRIVILEGE 1>, <PRIVILEGE 2>, <PRIVILEGE 3> ON VIEW <VIEW NAME> TO
ROLE <ROLE NAME>;
```

Multiple privileges can be granted at a time; however, privileges must be granted to only one role at a time. As with schema-level privileges, I highly recommend creating a script that generates SQL so that grants are consistently applied to all objects.

How Does Table Level Work with Schema Level?

Schema-level access and table-level access work well together because it allows an organization to be tactical about granular access on some assets, while painting broader strokes on some assets. This way, each object gets the treatment it requires, while the administration isn't overwhelmingly complex.

Organizations may decide that the majority of their data can be covered through schema-level access, but that certain datasets require greater protection, and therefore should be managed at the table or view level. In this situation, certain schemas should utilize schema-level access, and then designated schemas should use table-level access. Naming or documentation, preferably both, should indicate that a particular schema uses table-level grants instead of schema level.

For schemas using table-level access control, it is important that they are not granted the same privileges as the schema-level schemas. For example, bulk grants, specifically future grants, should not be utilized if a schema will use table-level access. Any access granted on a table after it is created would be in addition to the privileges implicitly granted through the future grants at table creation.

Key Takeaways

When working with privileges at the schema-object level, specifically with tables and views, the amount of work will be sufficiently higher than when working with schema-level privileges alone. As a result, we should be careful, logical, and consistent so that our grants work at scale and we do not skip objects. In the next chapters, we will cover granular access control. The following are the key takeaways from this chapter:

- Table- and view-level access control requires more administrative work than schema-level access because it involves more grants.

- Table- and view-level access can be used in addition to schema-level access by separating data into different schema types.

- Table- and view-level access can work well for special limited scope projects or users like a third-party tool or contractor.

146

CHAPTER 13

Row-Level Permissioning and Fine-Grained Access Control

Over the past few chapters, we've covered access control at the account, database, schema, and table levels. Account, database, and schema levels all involve permissions for a container rather than primary objects. This means controlling access to multiple objects at a time. In the last chapter, we covered the table level, the first object-level permissioning we've worked with. This meant allocating privileges on an object-by-object basis. In this chapter, we're going to start allocating privileges on a sub-object basis, meaning that the data we're protecting is a subset of data within an object. This is also when permissioning starts to get a bit trickier. In Figure 13-1, we can see where row-level privileges exist in our object hierarchy.

147

© Jessica Megan Larson 2022
J. M. Larson, *Snowflake Access Control*, https://doi.org/10.1007/978-1-4842-8038-6_13

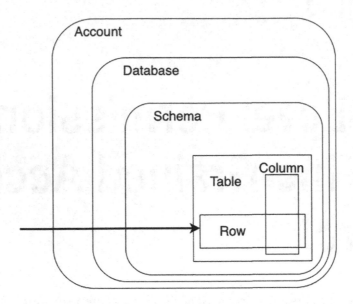

Figure 13-1. *The levels of access control in Snowflake*

In this chapter, we're going to cover row-level permissioning, also referred to as row-level filtering. We're going to address what row-level permissioning is, when it is beneficial, and a few different methods we can use to implement it. Since row level is a bit more nuanced than the other levels of access control, we will wrap up this chapter with some tips and tricks we can use to make implementing row-level access control easier.

What Is Row-Level Permissioning?

Row-level permissioning is when we allocate access to individual rows rather than an entire table or view based on some characteristic of the underlying data. We can allocate different access to different users on the same table. Row-level permissioning allows us to filter rows so that each user is able to view the subset of rows that they're allowed to view.

When we create row-level permissions, we don't individually grant access to each row in a table, we use logic on top of one or more characteristics of a row, combined with a characteristic or role of the user. This is important because most tables do not hold static data, the data changes over time, and therefore, directly granting access to rows would be an impossible task if it were an option. Additionally, it would only be possible for very small tables. Instead, we create mappings or rules to programmatically determine which users should have access to which rows of data. These mappings and rules combine the data from the row with characteristics of the user so that each user only sees the data they're allowed to view.

Row-level permissioning allows us to permission a table such that each individual or team sees a different subset of rows in a table. We can do that through mappings or through rules. In Figure 13-2, we can see how row-level permissions filter a table so that users with different roles can see different rows.

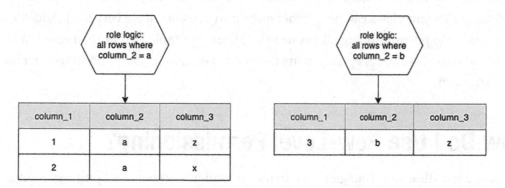

Figure 13-2. *An example of row-level access control filtering a table for different roles*

The first role has the privilege to see all rows where `column_2` has the value `'a'`. The second role has the privilege to see all rows where `column_2` has the value `'b'`. In this scenario, we're using very basic logic to segment the rows into different views for different roles.

Why Use Row-Level Permissioning?

Many organizations have the need to filter the output of a view or table so that different users see a different subset of rows. This may be required for datasets with particularly sensitive data. Row-level permissioning allows organizations to store all relevant data in a single table, rather than many tables with identical structure.

Organizations often have sensitive datasets where different users need a small subset of rows from a particular table or view, rather than the entire dataset. Since these datasets contain sensitive information, it is important to show users only the information they need. One example of this might be a table of contacts for the sales team. In this example, sales users should only be able to see the contacts for the organizations within their sales region. If we use row-level permissioning, we can accomplish this using a mapping table, or through logic.

Using row-level access control instead of simply segmenting each subset of rows into its own table or view benefits an organization in many ways. Keeping track of multiple tables and views can be cumbersome and does not scale very well. Additionally, using row-level permissioning allows us to use the same table in reports through a visualization tool, seamlessly displaying the correct data for each different user viewing the dashboard.

How Do I Use Row-Level Permissioning?

We have a few different strategies we can use to implement row-level permissioning. We can use secure views, meaning views that respect the RBAC boundaries, rather than regular views, which use the privileges of the creator of the view, and which may require that end users have access to the underlying data. We can also use row access policies, a feature added more recently by Snowflake. In either of these two methods, we can use logic and rules to create our row-level access control, or we can use a mapping table that holds a value about the user or role that can view something, in addition to a field that maps to a field in the target table.

Row Access Policies vs. Secure Views

End users will have the same experience querying a table with a row access policy as they will querying from a secure view. That is, if the same row-level permissions are added, the end user will have the same result. When choosing between using a row access policy and a secure view, one main concern is whether or not an account is configured for a row access policy. This is a feature that was released in the summer of 2021 and is available to Enterprise, Business Critical, and Virtual Private Snowflake account editions; however, it is not available in the Standard edition of Snowflake. To find out which edition of Snowflake your organization uses, you may need to reach out to your Snowflake representative, or support@snowflake.net.

Row access policies may be safer because they are actually placed on top of a table or view, whereas a secure view is a separate, duplicate view. When using a secure view, the original table must be protected in other ways, such as through schema- or table-level access. Some organizations may prefer row access policies for this reason. However, some organizations may prefer using secure views because it can be done fairly easily by analysts who may otherwise create views using SQL during a typical day.

Creating Row Access Policies

We can create row access policies to control row-level permissions on tables or views. We can easily create very simple policies with basic logic using only a few lines of SQL. Mapping tables allow us to be more dynamic about controlling row-level access and can be easier to create. We can also create complex row access policies with multiple input columns. We need to be careful about how we modify an access policy that is currently in use on a table or view. I recommend creating a Python script, or a Snowflake procedure to generate and optionally run the SQL to create these policies so that they can be deployed quickly using basic information provided by stakeholders. I prefer using Python because that is what I am more familiar with, and it fits my organization's practices, but it is important to point out that your organization may prefer to have everything in-platform, which makes procedures a better choice.

Creating a Basic Row Access Policy

When we create and apply a row access policy, there are a few steps we will need to do no matter how simple or complex the policy is. First, we need to create the policy. In order to create a policy, we will need the CREATE POLICY privilege on a schema. To create the policy, we simply run a query like

```
CREATE ROW ACCESS POLICY <POLICY NAME> AS (<INPUT PARAMETER>, <INPUT TYPE>)
-> <LOGICAL STATEMENT>;
```

Where policy name is a fully qualified name for the policy, input parameter is an alias for the column to use the policy on, input type is the type of that column, and logical statement is a statement that returns a Boolean value like CURRENT_ROLE() = 'SALES_WRITE'. We can match on the input parameter, which we can see in the following example:

```
CREATE ROW ACCESS POLICY ACCOUNT_OBJECTS.ACCESS_POLICIES.SALES_NUMBERS_ROW_
ACCESS_POLICY AS (ACCOUNT_EXECUTIVE VARCHAR) -> CURRENT_USER() = ACCOUNT_
EXECUTIVE;
```

This policy will only return rows where the current user is the account executive listed in the table or view. Figure 13-3 shows an example of our table.

sales_account_executive	q1_sales
jdoe	400
jsmith	355

Figure 13-3. *The PROD.SALES.SALES_NUMBERS table with data about the first quarter sales volume*

We can see the table in Figure 13-3 has two sales users, Jdoe and Jsmith, and their respective sales during the first quarter. Now that we have a row access policy, we can apply this policy to an existing table or view, or we can create a table or view and attach the policy at the time of creation. When adding a policy to an existing table or view, we must have the APPLY ROW ACCESS POLICY privilege on the account as well as OWNERSHIP on the table or view. To add the policy to an existing table, we can run the following query:

```
ALTER TABLE <TABLE NAME> ADD ROW ACCESS POLICY <POLICY NAME> ON (<INPUT
COLUMN>);
```

To add the policy we just created to our existing table, we can run the following statement:

```
ALTER TABLE PROD.SALES.SALES_NUMBERS ADD ROW ACCESS POLICY ACCOUNT_OBJECTS.
ACCESS_POLICIES.SALES_NUMBERS_ROW_ACCESS_POLICY ON (SALES_ACCOUNT_
EXECUTIVE);
```

It is important to note that the name of the input column does not need to match the name of the parameter in the access policy since the access policy has a different scope for variable names, though using the same naming for the columns and the parameters may simplify maintenance. If we wanted to add this policy to the table at creation time, we can do so like

```
CREATE TABLE PROD.SALES.SALES_NUMBERS (
  SALES_ACCOUNT_EXECUTIVE VARCHAR,
  Q1_SALES INT
) WITH ROW ACCESS POLICY ACCOUNT_OBJECTS.ACCESS_POLICIES.SALES_NUMBERS_ROW_
ACCESS_POLICY ON (SALES_ACCOUNT_EXECUTIVE);
```

It does not really matter whether we add the policy at the time of table creation, or if we add it later, as long as we are mindful of potential data exposure while the table does not have a policy applied. Revisiting our table, we can visualize the results of running `SELECT * FROM PROD.SALES.SALES_NUMBERS;` for a few users as shown in Figure 13-4.

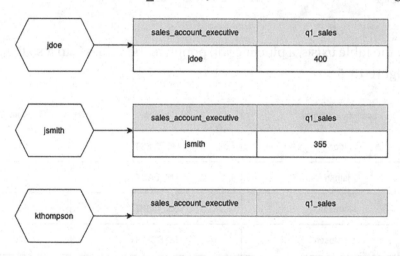

Figure 13-4. *The result of different users querying PROD.SALES.SALES_ NUMBERS after the row access policy has been applied*

As we can see, since the row access policy only shows rows where the querying user is the `sales_account_executive`, Jdoe can see one row, Jsmith can see one row, and Kthompson cannot see any rows since they are not the `sales_account_executive` on any rows in the source table.

Creating a Row Access Policy Using a Mapping Table

We can also use a mapping table to create and manage our row access policies. There are a few advantages of using a table to manage a row access policy, as well as a few drawbacks. The process of creating a policy using a mapping table is the same as the process for creating a basic policy, with the addition of creating the mapping table.

When we create a mapping table to use in a row access policy, we are substituting logical statements with mapping characteristics of the current user to characteristics about rows in the table the policy will be applied to. Let's walk through an example. We have a table for sales quotas, PROD.SALES.SALES_QUOTAS. This table tracks the sales quota for each account executive.

```
CREATE TABLE PROD.SALES.SALES_QUOTAS (
  ACCOUNT_EXECUTIVE VARCHAR,
  REGION VARCHAR,
  QUOTA NUMERIC
);
```

Now that the table exists, a pipeline will populate the values into it so that we have the data in Figure 13-5.

account_executive	region	quota
jdoe	US_WEST	550
jsmith	US_EAST	220
kjohnson	US_WEST	600
kthompson	US_SOUTH	375

Figure 13-5. *The table PROD.SALES.SALES_QUOTAS with account executives, their region, and their quotas*

We can see that each account executive has their username mapped to their region, and their respective quotas. Now, we can create a table that maps the managers to their account executives so that managers can see the quotas for their team members, but not for account executives on different teams.

```
CREATE TABLE PROD.SALES.AE_MGR_MAPPING (
  ACCOUNT_EXECUTIVE VARCHAR,
  SALES_MANAGER VARCHAR
);
```

This data will be populated by a pipeline from our employee management system so that the data is always up to date. We can see our mapping table looks as shown in Figure 13-6.

account_executive	sales_manager
jdoe	vlawrence
jsmith	sadams
kjohnson	vlawrence
kthompson	skerr

Figure 13-6. *The mapping table for PROD.SALES.SALES_QUOTAS that maps sales managers to account executives on their team*

As we can see in the two tables, the account_executive column matches across both. Again, these columns do not need to have the same name. We can now create a policy that maps the values from the mapping table to an input column like the following:

```
CREATE ROW ACCESS POLICY ACCOUNT_OBJECTS.ACCESS_POLICIES.AE_MGR_MAPPING AS
(AE VARCHAR) RETURNS BOOLEAN ->
EXISTS (
  SELECT TRUE
  FROM PROD.SALES.AE_MGR_MAPPING
  WHERE AE = ACCOUNT_EXECUTIVE
  AND SALES_MANAGER = CURRENT_USER()
);
```

Now that we've created the policy, we can apply it to the original table.

```
ALTER TABLE PROD.SALES.SALES_QUOTAS ADD ROW ACCESS POLICY ACCOUNT_OBJECTS.
ACCESS_POLICIES.AE_MGR_MAPPING ON (ACCOUNT_EXECUTIVE);
```

We've now applied the access policy to our table. Now we can visualize what the data would look like for our different users. The table in Figure 13-7 shows the result of running `SELECT * FROM PROD.SALES.SALES_QUOTAS` for each of the following users.

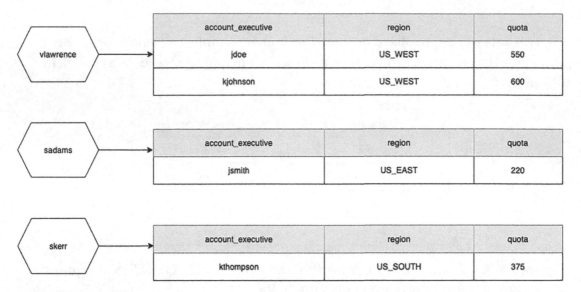

Figure 13-7. *Sales managers can see the data from their direct reports*

The user Vlawrence manages Jdoe and Kjohnson, and therefore can see those two rows from the table. Sadams manages Jsmith and can see that row, and Skerr manages Kthompson, and can see that single row.

Since we're managing this row access policy using a table that is populated from an employee management system, when the chain of management changes, the resulting privileges on the target table change as well. This has the benefit of constantly evolving to meet the needs of the business without requiring a Snowflake administrator to make changes each time it updates. This may be desirable in situations like this with reporting changes; however, it may not be ideal for more sensitive data like that subject to SOX controls.

Creating More Complex Row Access Policies

Since row access policies use Boolean SQL statements to determine access, we can make these policies as simple or as complex as we would like. We can use a hybrid approach of using a mapping table with additional logic. We can also create policies with multiple input parameters. Some complex queries may error because there is a limitation to how computationally intensive they can be, though Snowflake does not provide guidance on what that limit is. It is important to note that row access policies do require extra work to process, and therefore the more complex a policy, the more performance degradation we will see. We're going to run through a few examples of more complex policies.

In the first example, we're going to create a policy that allows a certain set of users to see all rows in the table, in addition to the sales managers from the previous section. In our fictional organization, the head of sales should be able to see all rows, regardless of who the manager is. We'll grant them the role SALES_EXECS and create the policy through the following statement:

```
CREATE ROW ACCESS POLICY ACCOUNT_OBJECTS.ACCESS_POLICIES.AE_MGR_MAPPING AS
(AE VARCHAR) RETURNS BOOLEAN ->
CURRENT_ROLE() = 'SALES_EXECS'
OR EXISTS (
  SELECT 1 FROM PROD.SALES.AE_MGR_MAPPING
  WHERE AE = ACCOUNT_EXECUTIVE
  AND SALES_MANAGER = CURRENT_USER()
);
```

Since we chained the Boolean statements together using OR, only one of the statements needs to evaluate to true for the row to be present.

Now we can imagine that we need to support an intermediary level of sales management, the sales region directors. In this example, we're going to assume that we have the following roles, SALES_DIRECTORS that every director in sales has access to, and a series of region roles that everyone within a region have access to. To accomplish this, we will add some logic on top of the last access policy, in addition to another parameter. Since this uses multiple rows, we're going to need to use the CURRENT_AVAILABLE_ROLES function. This gets a bit messy; we'll address strategies to clean this up in a later section of this chapter.

```
ARRAY_CONTAINS('SALES_' || REGION::VARIANT, PARSE_JSON(CURRENT_AVAILABLE_
ROLES())::ARRAY) AND ARRAY_CONTAINS('SALES_DIRECTOR'::VARIANT, PARSE_
JSON(CURRENT_AVAILABLE_ROLES())::ARRAY)
```

The preceding statement checks whether a user has access to the sales region role as well as the sales director role. Since this uses an AND connector, we need to be careful to put this in parenthesis within the row access policy. We will additionally need to add the region column to the input parameters.

```
CREATE ROW ACCESS POLICY ACCOUNT_OBJECTS.ACCESS_POLICIES.AE_MGR_MAPPING AS
(AE VARCHAR, REGION VARCHAR) RETURNS BOOLEAN ->
```

Now we can put all of this together into our policy.

```
CREATE ROW ACCESS POLICY ACCOUNT_OBJECTS.ACCESS_POLICIES.AE_MGR_MAPPING AS(
  AE VARCHAR, REGION VARCHAR)
RETURNS BOOLEAN ->
CURRENT_ROLE = 'SALES_EXECS'
OR (
  ARRAY_CONTAINS('SALES_' || REGION::VARIANT,
  PARSE_JSON(CURRENT_AVAILABLE_ROLES())::ARRAY) AND
  ARRAY_CONTAINS('SALES_DIRECTOR'::VARIANT,
  PARSE_JSON(CURRENT_AVAILABLE_ROLES())::ARRAY)
)
OR EXISTS (
  SELECT 1 FROM PROD.SALES.AE_MGR_MAPPING
  WHERE AE = ACCOUNT_EXECUTIVE
  AND SALES_MANAGER = CURRENT_USER()
);
```

Since we've added a parameter, we will need to change how we apply this to
the table.

```
ALTER TABLE PROD.SALES.SALES_QUOTAS ADD ROW ACCESS POLICY ACCOUNT_OBJECTS.
ACCESS_POLICIES.AE_MGR_MAPPING ON (ACCOUNT_EXECUTIVE, REGION);
```

As with all functions, the ordering of the columns is important, the ordering when applying
to the table must match the ordering on the row access policy for it to work as desired. If we
create a policy that errors because it is too computationally complex, we can move some of the
logic out of the policy and into a pre-computed mapping table, or into a UDF.

Removing a Row Access Policy

To delete a row access policy, we must first remove it from the table or view, and then
we can drop the policy entirely. If we do not remove the policy, dropping the policy will
error, letting the user know that the policy is associated with one or more tables or views.
To remove a row access policy, we can run the following statement:

```
ALTER TABLE <TABLE NAME> DROP ROW ACCESS POLICY <ROW ACCESS POLICY NAME>;
```

Now that we've detached the policy from the table, we can drop the policy definition itself.

```
DROP ROW ACCESS POLICY <ROW ACCESS POLICY NAME>;
```

The syntax is the same as dropping any other database object. Our policy now does not exist on the table and has been cleaned up from the database.

Altering Row Access Policies

When we need to modify or remove a row access policy from a table or view, there are some constraints that we will need to work with. We cannot drop a row access policy that is currently attached to a table. We can, however, alter an existing row access policy while it is applied to a table.

If we want to change the definition of a row access policy, we can start by querying the definition. We can do this by running the following statement:

```
SELECT GET_DDL('POLICY', <ROW ACCESS POLICY NAME>);
```

This will return the definition of the policy, which we can use to draft a new policy, or to make modifications. If we're unsure which column or columns we use as input for a policy, we can also run

```
DESCRIBE ROW ACCESS POLICY <ROW ACCESS POLICY NAME>;
```

This command does not give us a lot of information, but may be helpful to know in certain situations.

When we modify an existing access policy, I recommend using an alter statement rather than dropping and recreating or using a CREATE OR REPLACE clause. This is because we cannot drop a policy that is attached to a table. When we replace a policy, that is the same operations as dropping the policy and creating a new one, and if the policy is attached to a table or view, the operation will err. Fortunately, the syntax for altering a policy is very similar to that of creating it. To alter a policy, run the following statement:

```
ALTER ROW ACCESS POLICY <ROW ACCESS POLICY> SET BODY -> <EXPRESSION>;
```

CHAPTER 13   ROW-LEVEL PERMISSIONING AND FINE-GRAINED ACCESS CONTROL

We can see this in practice in the following example. Here we're going to reset the definition of our policy ACCOUNT_OBJECTS.ACCESS_POLICIES.AE_MGR_MAPPING to simply check if the current role is SALES_EXECS.

```
ALTER ROW ACCESS POLICY ACCOUNT_OBJECTS.ACCESS_POLICIES.AE_MGR_MAPPING SET
BODY -> CURRENT_ROLE = 'SALES_EXECS';
```

Since this simply modifies the existing policy in place, the table or view will immediately reflect this change.

Creating a Python Script to Generate Row Access Policies

Since row access policies require a bit of SQL and are on a table-by-table basis, I recommend generalizing the access patterns and creating a Python script to generate the SQL, and optionally running the SQL directly from that script. Since there are two main ways of creating a row access policy, I recommend creating both of those pathways.

First, we define the parameterized strings for creating the row access policy, and for applying the policy to the table. A common pattern I anticipate is having a mapping table, in addition to one or more roles that will need access to all rows all the time.

```
ROW_ACCESS_POLICY_SQL = """
CREATE ROW ACCESS POLICY ACCOUNT_OBJECTS.ACCESS_POLICIES.{table_name} AS
({input_cols_typed}) RETURNS BOOLEAN ->
{allowed_roles_clause}
{optional_or}
EXISTS (
  SELECT 1 FROM {mapping_table}
  WHERE {column_mappings_sql});
"""
```

Next, the parameterized string for applying the policy to a table.

```
APPLY_POLICY_SQL = """
ALTER TABLE {table_name}
ADD ROW ACCESS POLICY ACCOUNT_OBJECTS.ACCESS_POLICIES.{table_name}
ON ({input_cols_named});
"""
```

Now we can define the inputs to our function and create our function doc string. It's especially important to be very clear here since we may use this function with manual input.

```
def create_row_access_policy(
        table_name,
        input_columns,
        mapping_table,
        column_mappings,
        allowed_roles,
    ):
    """
    :param table_name: the fully qualified table name
    :type table_name: str
    :param input_columns: the arguments to the policy in form [{'name':
     col_name, 'type': col_type},...]
    :type input_columns: list
    :param mapping_table: the fully qualified table name for mapping values
    :type mapping_table: str
    :param column_mappings: mapping column names in the target table to
     the matching columns in the mapping table in form [(target_col1,
     mapping_col1),]
    :type column_mappings: list
    :param allowed_roles: roles that should have access to all rows
     regardless of logic
    :type allowed_roles: list
    """
```

Now that we have our inputs, we can start to transform them to fit the structure we will need them in for the SQL statements.

```
input_cols_named = ', '.join(x['name'] for x in input_columns)
input_cols_typed = ', '.join([x['name'] + ' ' + x['type'] for x in input_
columns])
column_mappings_sql = '\nAND '.join([x[0] + ' = ' + x[1] for x in column_
mappings])
allowed_roles_clause = '\nOR '.join(["CURRENT_ROLE() = '{r}'".format(r=x)
for x in allowed_roles])
```

Now that we've done the majority of the transformation work, we can pass these formatted strings into our original parameterized strings.

```
access_policy_sql = ROW_ACCESS_POLICY_SQL.format(
    table_name=table_name,
    input_cols_typed=input_cols_typed,
    allowed_roles_clause=allowed_roles_clause,
    optional_or='OR ' if allowed_roles_clause else '',
    mapping_table=mapping_table,
    column_mappings_sql=column_mappings_sql,
)
apply_sql = APPLY_POLICY_SQL.format(
    table_name=table_name,
    input_cols_named=input_cols_named,
)
```

Finally, we can return the SQL we generated, and optionally run it directly on Snowflake, write it to a DDL file to check into version control, or any other use we have for it.

```
return access_policy_sql, apply_sql
```

This script will be included in the source code in its entirety, alongside other scripts. We can use a similar method to create a script for other types of row access policies too.

Constraints

We have a few limitations and constraints when creating row access policies. When using a mapping table, we cannot also use the function IS_ROLE_IN_SESSION. In this situation, it may be better to rely on functions like CURRENT_ROLE or ALL_AVAILABLE_ROLES. Another major thing to keep in mind is that if we have a service account populating a table in a way other than append-only, we need to make sure that account can view all of the rows in the table. This is because many operations require read, such as updating values and deduping a table.

Using Secure Views

We can use secure views to accomplish a similar end user experience as row access policies with row-level access control. Row access policies directly filter the data in tables in views, when we use views instead, we need to make sure that we're adequately protecting the source data since we will be essentially duplicating the data when we pull from one or more tables. To do this, I recommend storing the source tables in a raw schema with a very small audience. It is important to note that when we use this method, we cannot prevent the creator of these views from viewing the sensitive source data, which may mean that row access policies are a better fit for more sensitive data.

Before we walk through an example using secure views, let's cover what exactly a secure view is. With a standard view, users may need access to the underlying data in order to query data in the view. With a secure view, that is not the case. The underlying data stays protected. A secure view also hides the definition SQL from users who do not have the ownership privilege. We can see how we might separate the source table from the secure view using separate schemas for a user in Figure 13-8.

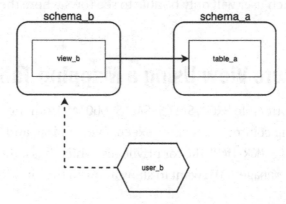

Figure 13-8.   *A secure view pulls data from a table in schema_a, which is not accessible to user_b, however the view in schema_b is*

When we create a secure view for row-level access control, we are essentially drafting one or more Boolean statements that, when evaluated as true, will allow the row to appear. These statements will appear as part of the where clause in the view creation. We're going to revisit the same problems we solved in the previous section, but this time using secure views.

Creating a Basic Secure View

In the first example, we have the table PROD.SALES.SALES_NUMBERS, which has the following columns: sales_account_executive, and q1_sales. We want to create a secure view that only shows a row to a user if that user is the sales account executive. We can do this with a secure view by using the Boolean statement CURRENT_USER() = SALES_ACCOUNT_EXECUTIVE in the where clause like

```
CREATE SECURE VIEW PROD.SALES_VIEWS.SALES_NUMBERS_VIEW AS (
SELECT *
FROM
PROD.SALES.SALES_NUMBERS
WHERE CURRENT_USER() = SALES_ACCOUNT_EXECUTIVE
);
```

In this situation, we're going to create the view in a different schema SALES_VIEWS which contains cleaned and secured data assets. Just as in the previous section using row access policies, each user will only be able to see rows where they're the account executive.

Creating a Secure View Using a Mapping Table

Next, we're revisiting our table PROD.SALES.SALES_QUOTAS from the previous section, which has the following columns: account_executive, region, and quota. We will be using PROD.SALES.AE_MGR_MAPPING for privileges, which has columns account_executive and sales_manager. We want to show a row in the view if the current user is the sales manager.

```
CREATE SECURE VIEW PROD.SALES_VIEWS.SALES_QUOTAS_VIEW AS (
SELECT S.*
FROM PROD.SALES.SALES_QUOTAS S
INNER JOIN PROD.SALES.AE_MGR_MAPPINGS M
ON S.ACCOUNT_EXECUTIVE = M.ACCOUNT_EXECUTIVE
WHERE CURRENT_USER() = M.SALES_MANAGER
);
```

Rather than comparing the data in the mapping table to an input variable, we're joining it to the original table. We can then add the logic of mapping the current user to the sales manager in the mapping table to the where clause.

Creating More Complex Secure Views

In this example, we want to allow the head of sales to see all rows, in addition to the logic we had in the previous example. The head of sales uses the role SALES_EXECS to view assets like this. Since we want to show a row in the results if the user has the SALES_EXECS role **or** they are the manager, we will need to update our logic from the previous view. If we use an inner join, this will prevent the head of sales from seeing any data. Instead, we will change this to a left join, and add the mapping logic to the where clause.

```
CREATE SECURE VIEW PROD.SALES_VIEWS.SALES_QUOTAS_VIEW AS (
SELECT S.* FROM
PROD.SALES.SALES_QUOTAS S
LEFT JOIN PROD.SALES.AE_MGR_MAPPINGS M
ON S.ACCOUNT_EXECUTIVE = M.ACCOUNT_EXECUTIVE
WHERE CURRENT_USER() = M.SALES_MANAGER
OR CURRENT_ROLE() = 'SALES_EXECS'
);
```

In the next example, we want to build on this mapping and additionally include the regional director. Each regional director should be able to see all rows that pertain to their region. In the previous section, we walked through the logic of checking the available roles, which we're going to use here as well.

```
CREATE SECURE VIEW PROD.SALES_VIEWS.SALES_QUOTAS_VIEW AS (
SELECT S.* FROM
PROD.SALES.SALES_QUOTAS S
LEFT JOIN PROD.SALES.AE_MGR_MAPPINGS M
ON S.ACCOUNT_EXECUTIVE = M.ACCOUNT_EXECUTIVE
WHERE (CURRENT_USER() = M.SALES_MANAGER
  OR CURRENT_ROLE() = 'SALES_EXECS'
  OR (
    ARRAY_CONTAINS('SALES_' || REGION::VARIANT,
    PARSE_JSON(CURRENT_AVAILABLE_ROLES())::ARRAY) AND
```

```
    ARRAY_CONTAINS('SALES_DIRECTOR'::VARIANT,
    PARSE_JSON(CURRENT_AVAILABLE_ROLES())::ARRAY))
));
```

We joined this clause with an OR in the where clause so that it will not affect the existing row permissioning, and only add the new role's permissions.

Creating a Python Script to Generate Secure Views

Just as we created a script to generate the SQL for row access policies, we can create a script to create secure views. However, as we saw earlier in this section, secure views are a little trickier to generate programmatically. This may not be as useful for your organization since clauses here can get quite complex. My organization found a need for this as a way to automatically generate secure views from input data, so I have included this section as a result.

Just as we defined a parameterized string to start our row access policy Python script, we can do that here.

```
SECURE_VIEW_SQL = """
CREATE SECURE VIEW {database}.{schema}.{view} AS
(SELECT * FROM {table}
    WHERE
    {clauses});
"""
```

Next, we'll define our inputs. For this, I'm creating a function that will accept logical clauses and a list of allowed roles. This can be used by wrapper functions that create the logic from inputs if desired.

```
def create_secure_view(
    database,
    schema,
    view,
      table_name,
    logical_statements,
      allowed_roles=[],
    ):
```

```
"""
:param database: the database for the view
:type database: str
:param schema: the schema for the view
:type schema: str
:param view: the name of the view
:type view: str
:param table_name: the fully qualified table name
:type table_name: str
:param logical_statements: a series of logical statements to be joined
 together with OR
:type logical_statements: list
:param allowed_roles: roles that should have access to all rows
 regardless of logic
:type allowed_roles: list
"""
```

Now we can start the transformation logic. Since this is a very basic function, there is minimal transformation.

```
allowed_roles_sql = '\nOR '.join(['CURRENT_ROLE() = ' + x for x
    in allowed_roles])
clauses = '\n OR '.join(logical_statements)
if allowed_roles_sql:
    clauses += '\nOR ' + allowed_roles_sql
```

Finally, we can put our transformed clauses into our parameterized string.

```
secure_view_sql = SECURE_VIEW_SQL.format(
    database=database,
    schema=schema,
    view=view,
    table=table_name,
    clauses=clauses,
)
```

And that's it! Just like with the previous script, this will be available in its entirety alongside other scripts.

Tips and Tricks

In addition to the strategies discussed in previous sections, there are a number of things we can do to make the process of managing row-level permissions easier. We can create some *user defined functions* (UDFs) to simplify the queries used in row access policies or secure views. Aggregate statistics will reflect the rows a user has access to and may not be accurate. When using row-level permissioning with secondary roles, we may need to utilize different functions for checking privileges.

User-Defined Functions

Creating a few UDFs to match common access patterns can greatly simplify the process of creating row access policies or secure views. I recommend creating a simple UDF that checks whether or not a user has a particular role at their disposal. This can be called from a secure view, or a row access policy.

```
CREATE OR REPLACE FUNCTION IS_AVAILABLE_ROLE(role_name varchar)
RETURNS BOOLEAN
LANGUAGE SQL
    AS $$
        ARRAY_CONTAINS(role_name::variant, parse_json(current_available_
        roles())::array)
    $$;
```

This UDF can have the input from a hard coded role, or from a value in a column. For example, with the region in a preceding example, an analyst could simplify the logic. The original logic looked like

```
ARRAY_CONTAINS('SALES_' || REGION::VARIANT,
  PARSE_JSON(CURRENT_AVAILABLE_ROLES())::ARRAY) AND
ARRAY_CONTAINS('SALES_DIRECTOR'::VARIANT,
  PARSE_JSON(CURRENT_AVAILABLE_ROLES())::ARRAY))
```

Using the new function, the logic becomes

```
IS_AVAILABLE_ROLE('SALES_' || REGION)
AND
IS_AVAILABLE_ROLE('SALES_DIRECTOR')
```

This is much more user-friendly than an analyst trying to work with system functions.

Aggregate Statistics

When querying a table with a row access policy, or a view with row-level permissioning, it is important to remember that aggregate statistics will reflect the subset of rows that a user or role has access to. If a table has 100 rows, 50 of which an analyst has privileges to view, when the analyst runs SELECT COUNT(*) FROM TEST_TABLE; the result will be 50, not 100. If they include a figure like this on a dashboard, it may show a different count for each user, which can create confusion. If aggregate statistics are necessary, it might be a good idea to create some aggregate tables or views upstream of the row-level data.

Row-Level Permissioning with Secondary Roles

When using secondary roles, users may not spend much time thinking about their primary role and may primarily use their default role. Because of this, we need to make sure that the ways we're writing row-level permissioning matches the way users interact with Snowflake. If your organization uses secondary roles, it may not make sense to use the function CURRENT_ROLE, and instead relying when you can on ALL_AVAILABLE_ROLES or IS_ROLE_IN_SESSION. As a reminder, IS_ROLE_IN_SESSION cannot be used in row access policies that use mapping tables. Otherwise, this function is appropriate to use.

Key Takeaways

As we saw in this chapter, we have many different methods at our disposal for implementing row-level access control. Your organization may choose to utilize one of these methods, or it may make more sense to deploy a hybrid solution. My organization has opted to use a hybrid approach where the most sensitive data is controlled with row access policies created by the data platform team, and less sensitive data can be protected by secure views created by analysts. As always, there is no single best solution. In the next chapter, we will continue working with granular access controls, shifting our focus to protecting columns rather than rows. The following are key takeaways from this chapter:

- Row-level permissioning filters a table or view so that each user sees a subset of the rows.

- We can create row-level permissioning through logic, a mapping table, or a combination of both.

- Row access policies are one way of creating row-level permissioning and they are directly applied to tables.

- Secure views pull data from a more restricted table or view and filter the rows so that different users can see a different subset of data.

- Secondary roles mean users may not utilize the appropriate primary role, which means that we need to use different logic to check access.

CHAPTER 14

Column-Level Permissioning and Data Masking

In the previous chapter, we covered row-level permissioning. This meant filtering the output from a table or view so that each user could see the subset of rows they're allowed to see. We could do this via row access policies, or using secure views combined with logic and/or a mapping table. In this chapter, we're going to cover column-level permissioning and data masking, which we can visualize in the graphic shown in Figure 14-1.

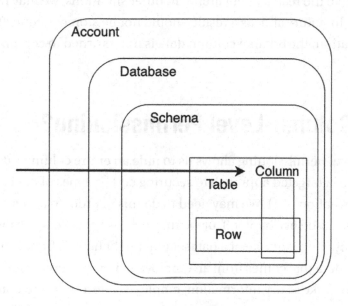

Figure 14-1. *The levels of access control in Snowflake*

© Jessica Megan Larson 2022
J. M. Larson, *Snowflake Access Control*, https://doi.org/10.1007/978-1-4842-8038-6_14

As you can see in the graphic in Figure 14-1, we've reached the final level of access control in the Snowflake object hierarchy. In this chapter, we're going to cover the different strategies we can use to create column-level access control and how we can manage it more easily. We'll discuss creating these permissions using dynamic masking policies, using secure views, and using external tokenization. Some of the strategies we will address will mirror strategies we employed during row-level access control, and some will be entirely new.

What Is Column-Level Permissioning?

Column-level permissioning is when we control which users can view a particular column in a table or view. Looking back at row-level permissioning, users will receive a subset of the rows in a table. Column-level permissioning, unlike with row-level permissioning, we cannot simply omit a column in a result because SQL does not permit that. When we use column-level permissioning, users will receive all of the rows; however, some of the data in columns will be replaced with null values, spoofed values, or tokenized values. In some situations, we may return null values as that may be sufficient for our use cases. However, in some situations, we may prefer spoofed values that look like the real deal but aren't. In other situations, we may have data that is extra-sensitive in nature and, as a result, should not be stored in Snowflake in its raw form. In any situation, the original column data is not returned to certain users in certain situations.

Why Use Column-Level Permissioning?

Since column-level permissioning allows us to hide an entire column's data, we can use this to employ a targeted approach to securing certain types of data like Personally Identifiable Information (PII). We may need to do this in order to adhere to regulatory requirements discussed in Chapter 3, or we may do this because we're just concerned about data privacy. One example of a dataset that might benefit from column-level permissioning is data that comes from an employee management system, including sensitive data types like salary information, employee addresses, and dates of birth.

We may decide that with this employee dataset, the head of HR can unmask salary data for auditing purposes, but that no one else should be able to see that data, including members of an HR team that may need to work with this dataset. The easiest way to handle this is to mask the salary column, and all other columns with sensitive data so that HR team members can interact with the tables and views without being exposed to information they do not need to see in order to do their job. This way they're still able to do their job with minimal hurdles, but the exposure risk of this data is low.

Using column-level permissioning protects our most sensitive datasets while still allowing users to get work done. This also allows us to be more precise and minimize our risk of data leakage vs. locking down entire tables or views from most users.

How Do I Implement Column-Level Permissioning?

There are three ways we can use to implement column-level permissioning: by using a dynamic masking policy, by creating a secure view, or by using external tokenization. External tokenization is the most secure form of column-level permissioning as the source data is never stored in its raw form in Snowflake, just tokenized data. Masking policies are the next most secure. Like row access policies, masking policies sit directly on top of the table or view, meaning that the source data is protected. Finally, there are secure views, which aren't as secure as the other two, but still provide the ability to mask columns. As we discussed in the last chapter, secure views essentially duplicate the data to another object meaning that both the original object, and the secure view must be protected. We can also combine masking policies and external tokenization to de-mask certain fields using external functions.

For creating a masking policy and a secure view, just like with row-level permissioning, we can create column-level privileges using logical statements, or by using a mapping table. This allows us to be flexible with our approach. In some situations, we may want the policy to be harder to amend. In other situations, we want the policy to be updated on demand as the mapping data changes. This way we can choose the right method to solve our problem. However, mapping tables are a much less common pattern with column-level access.

Using Dynamic Masking Policies

We can create dynamic masking policies in much the same way as we created row access policies. The process is the same, except the masking policy is applied to columns within tables and views rather than to entire tables and views. First, we create the policy, which will accept a single column at a time, and then we can apply it to as many columns in tables or views as we would like using ALTER TABLE statements. Since masking policies are column or field specific, we should create generic policies for different types of data. For example, we might want to mask emails and social security numbers differently, and so it might make the most sense to have an ssn_mask and an email_mask. Since some of these types of data appear in multiple locations, we can reuse the same mask on multiple columns. Since we anticipate reusing masks, we should take care to follow naming conventions, and place all of our masks in a dedicated schema.

We also need to make sure that the correct access is granted to our users. To create an access policy, a user needs the CREATE MASKING POLICY privilege on a schema. To apply a masking policy, the APPLY MASKING POLICY privilege must be granted at the account level; the role must also have APPLY on the masking policy itself in the case that the role applying the policy is not the same as the role creating the policy. And finally, the role must have OWNERSHIP on the table in order to modify it to include a masking policy on one or more columns.

Creating a Basic Masking Policy

We can create a masking policy to mask salary data, SALARY_MASK. This mask will replace salary data with a blank cell. We're going to use this mask on a table, PROD.EMPLOYEES. SALARIES, shown in Figure 14-2.

employee_name	job_title	salary
jdoe	account executive	$60,000
jsmith	account executive	$90,000
kjohnson	accountant	$85,000

Figure 14-2. *Unmasked data from PROD.EMPLOYEES.SALARIES with three employee salaries*

Before we can apply the policy to our table, we must create the mask, carefully, making sure that the input datatype is the same as the output datatype:

```
CREATE MASKING POLICY PROD.ACCESS_POLICIES.SALARY_MASK AS
(SALARY VARCHAR) RETURNS VARCHAR ->
CASE
WHEN CURRENT_ROLE() = 'HR_ADMIN' THEN SALARY
ELSE NULL
END;
```

We can now apply this policy to the salary column in our table PROD.EMPLOYEES. SALARIES, which maps employees to their respective salaries. Snowflake will only allow one masking policy to be applied on a column at a time, and if there is an attempt to apply a masking policy to a column that already has one set, the operation will error, leaving the initial masking policy intact.

```
ALTER TABLE PROD.EMPLOYEES.SALARIES MODIFY COLUMN SALARY SET MASKING POLICY
PROD.ACCESS_POLICIES.SALARY_MASK;
```

Our policy will allow users with the HR_ADMIN role to see all of the data in the table, and anyone querying the table with a different role will see the salary column masked. For a user without this role, the table will look as shown in Figure 14-3.

employee_name	job_title	salary
jdoe	account executive	
jsmith	account executive	
kjohnson	accountant	

Figure 14-3. *The result of a non-hr-admin user querying the masked table*

Let's say that instead of just having the policy return a null value, we instead want it to look like $XXX,XXX. We can unset the policy on the column, update the policy, and add the policy to the column again. We can also alter the policy in place. To unset, change, and then re-add, we can run the following:

```
ALTER TABLE PROD.EMPLOYEES.SALARIES MODIFY COLUMN SALARY UNSET
MASKING POLICY;
CREATE OR REPLACE MASKING POLICY PROD.ACCESS_POLICIES.SALARY_MASK AS
```

```
(SALARY VARCHAR) RETURNS VARCHAR ->
CASE
WHEN CURRENT_ROLE() = 'HR_ADMIN' THEN SALARY
ELSE '$XXX,XXX'
END;
ALTER TABLE PROD.EMPLOYEES.SALARIES MODIFY COLUMN SALARY SET MASKING POLICY
PROD.ACCESS_POLICIES.SALARY_MASK;
```

The benefits to doing this is that we could actually create a new policy, wrapping this operation in a transaction block, ensuring that our data is always protected. To alter the policy in place we can do the following:

```
ALTER MASKING POLICY PROD.ACCESS_POLICIES.SALARY_MASK SET BODY ->
CASE
WHEN CURRENT_ROLE() = 'HR_ADMIN' THEN SALARY
ELSE '$XXX,XXX'
END;
```

Now, when a non-hr-admin user views this table, they will see what is shown in Figure 14-4, while the view for a user with HR_ADMIN will remain the same with the original values.

employee_name	job_title	salary
jdoe	account executive	$XXX,XXX
jsmith	account executive	$XXX,XXX
kjohnson	accountant	$XXX,XXX

Figure 14-4. *The result of masking the data in the salaries table with more realistic looking data*

Sometimes replacing sensitive data with fake looking data, like in Figure 14-4, helps because it signals to users that the data was masked. We could easily replace this with something like '***REDACTED***' or with instructions on how to access this field if the data in question is originally a string.

If we shift gears to another dataset, like one with employee emails or phone numbers, we can selectively mask a column using regex. For example, if we have the dataset PROD.EMPLOYEES.CONTACTS with email and phone number data, as shown in Figure 14-5, we can partially mask values.

employee_name	email	phone
jdoe	jdoe@gmail.com	(555)555-1234
jsmith	embarassing_nickname@aol.com	(555)555-9876
kjohnson	k_johnson@berkeley.edu	(555)555-0000

Figure 14-5. *The unmasked PROD.EMPLOYEES.CONTACTS data with employee emails and phone numbers*

Let's create a mask, EMAIL_MASK that uses regex to select some of the original data and mask the rest.

```
CREATE MASKING POLICY PROD.ACCESS_POLICIES.EMAIL_MASK AS
(EMAIL VARCHAR) RETURNS VARCHAR ->
CASE WHEN CURRENT_ROLE = 'HR_ADMIN' THEN EMAIL
ELSE REGEXP_REPLACE(EMAIL, '.+\@', '*****@')
END;
```

We could also mask our phone column so that the area code is visible, but nothing else is. This mask assumes that all phone numbers are in the same format.

```
CREATE MASKING POLICY PROD.ACCESS_POLICIES.PHONE_MASK AS
(PHONE VARCHAR) RETURNS VARCHAR ->
CASE WHEN CURRENT_ROLE = 'HR_ADMIN' THEN PHONE
ELSE SUBSTR(PHONE, 0, 5) || '***-****'
END;
```

Once we've applied these policies, our dataset now looks as shown in Figure 14-6.

employee_name	email	phone
jdoe	*****@gmail.com	(555)***-****
jsmith	*****@aol.com	(555)***-****
kjohnson	*****@berkeley.edu	(555)***-****

Figure 14-6. *The result of partially masking the email and phone columns in PROD.EMPLOYEES.CONTACTS*

Masking policies are flexible and can be used to create more complicated logic to solve more complicated problems, while at the same time, they're easy to create and apply.

Creating a Python Script to Generate Masking Policies

We can easily create a Python script to generate masking policies so that we can quickly, easily, and consistently create masking policies. Just like with the script we wrote for generating row access policies, we want to define parameterized SQL for the creation of the policy, as well as for applying the policy.

```
MASKING_POLICY_SQL = """
CREATE MASKING POLICY PROD.ACCESS_POLICIES.{field_name}_MASK AS
({field_name} {field_type}) RETURNS {field_type} ->
CASE
{body}
END;
"""

APPLY_POLICY_SQL = """
ALTER TABLE {table_name}
MODIFY COLUMN {column_name}
SET MASKING POLICY PROD.ACCESS_POLICIES.{field_name};
"""
```

We can additionally define a helper SQL block to help us populate the case when statement using the CURRENT_ROLE function. Another option is to use the IS_GRANTED_ TO_INVOKER_ROLE function, which supports role inheritance.

```
WHEN_CLAUSE_SQL = "WHEN CURRENT_ROLE() = '{role}' THEN {return_val}"
```

Now we can define the inputs for the function now that we have an idea of what the SQL requires. We'll also include a nice docstring that explains what the input data should look like.

```
def create_column_masking_policy(
    table_name,
    field_name,
    field_type,
    columns,
    role_mappings,
    allowed_roles,
    else_value,
    ):
    """

    :param table_name: the fully qualified table name
    :type table_name: str
    :param field_name: the type of value i.e., 'phone' or 'email'
    :type field_name: str
    :param field_type: the snowflake type of the field
    :type field_name: str
    :param columns: the columns to apply the policy to
    :type columns: list
    :param role_mappings: roles and their return value. i.e., [{role:
    'HR_ADMIN', 'return_val': 'phone'}]
    :type role_mappings: list
    :param allowed_roles: roles that should always see the unmasked column
    :type allowed_roles: list
    :param else_value: the value that should be returned if not an
    allowed role
    :type else_value: string
    """
```

In this function, we're going to apply masking policies on a table-by-table basis, but we could just as easily change the way we input functions and have the column parameter included as a list of columns per table. Now we can start to transform the input.

```
for role in allowed_roles:
    role_mappings.append({
        'role': role,
        'return_val': field_name})
when_clauses = '\n'.join([WHEN_CLAUSE_SQL.format(
        role=x['role'],
        return_val=x['return_val']) for x in role_mappings])
else_case = 'ELSE ' + else_value
body = when_clauses + '\n' + else_case
```

Here, we're just adding the allowed roles to the logical statements list so that we can process those in the same way we do the other logical statements. We then add the else value to the body of the policy. We can put this all together into our masking policy creation SQL.

```
mask_sql = MASKING_POLICY_SQL.format(
    field_name=field_name,
    field_type=field_type,
    body=body,
)
```

To apply the policy to each of our columns, we will iterate through each of the columns in the list and add the resulting SQL to a list.

```
apply_sql = []
for column in columns:
    apply_sql.append(APPLY_POLICY_SQL.format(
        table_name=table_name,
        column_name=column,
        field_name=field_name,
    ))
```

Now we just need to return our SQL.

```
return mask_sql, '\n'.join(apply_sql)
```

That's it! This script will be included alongside other scripts from this book.

Masking JSON Data

We can also use masking policies to mask data in JSON columns. To do this, we need to do a few things. First, we need to make sure that the JSON data in our columns has been parsed rather than being a string. Then we will create a JavaScript UDF that we will call in our mask definition. We will follow the same steps to create and apply our mask.

Suppose we have a table, PROD.TEST.JSON_DATA that has one column value with a string containing JSON data. The JSON data has three keys, name, age, and phone_number. We want to create a mask for phone_number. First, we need to massage the data so that it is in the right format.

```
CREATE TABLE PROD.TEST.PARSED_JSON_DATA AS (
  SELECT PARSE_JSON(VALUE) AS VALUE
  FROM JSON_DATA
);
```

Next, we can create our JavaScript UDF that will mask phone_number.

```
CREATE FUNCTION PROD.FUNCTIONS.PHONE_NUMBER_JSON_MASK (V VARIANT)
RETURNS VARIANT
LANGUAGE JAVASCRIPT
AS
$$
if ("phone_number" in V) {
  V["phone_number"] = "**REDACTED**"
}
return V;
$$;
```

We now have a function we can create our mask with.

```
CREATE MASKING POLICY PROD.ACCESS_POLICIES.PHONE_NUMBER_JSON_MASK AS
(VALUE VARIANT) RETURNS VARIANT ->
CASE WHEN CURRENT_ROLE = 'ADMIN' THEN VALUE
ELSE PROD.FUNCTIONS.PHONE_NUMBER_JSON_MASK(VALUE)
END;
```

We then apply this mask to the column with JSON data.

```
ALTER TABLE PROD.TEST.PARSED_JSON_DATA ALTER COLUMN VALUE SET MASKING
POLICY PROD.ACCESS_POLICIES.PHONE_NUMBER_JSON_MASK;
```

And now we're done. Our JSON will now show `'**REDACTED**'` instead of any phone number data.

Constraints

When pulling data from a table with column-level security, beware that any tables created using this data will respect the level of access that the creating role has. This means that if the role creating the downstream table has access to all columns unmasked, the resulting table will have all of the unmasked data. In situations like this, it is recommended to mask the resulting columns as well. The reverse can also be true, if a role does not have access to the unmasked data and creates a downstream table, that table will have masked data in it rather than the unmasked data.

Using Secure Views

We can create column-level access control using secure views, as we did with row-level access control. When we created these views for row-level access control, we used statements in the WHERE clause to filter the set of rows users have access to. For column level, we need to manipulate the way we select columns. We will heavily employ CASE statements in the select clause to accomplish this.

Remember that since we're using secure views to handle this, we need to make sure that we're adequately protecting the source data in addition to the data pulled into the secure view.

Creating a Basic Secure View

We're going to walk through the same examples we went through in the last section for masking policies, but this time, we will use secure views to solve these problems. If we recall, secure views hide the definition of the view, as well as prevent certain optimizations that could expose data to a user, which makes them ideal for using for access control. We have salary data in a table called PROD.EMPLOYEES.SALARIES. This table has the columns employee_name, job_title, and salary. We're going to mask the salary column so that only users with the HR_ADMIN role can view the unmasked data.

```
CREATE SECURE VIEW PROD.EMPLOYEES_CLEANED.SALARIES AS (
SELECT
EMPLOYEE_NAME,
JOB_TITLE,
CASE WHEN CURRENT_ROLE() = 'HR_ADMIN' THEN SALARY
ELSE NULL
END AS SALARY_MASKED
FROM PROD.EMPLOYEES.SALARIES
);
```

In this instance, I'm renaming the column to salary_masked as well to signal to users that this data is masked and that they may not see the source data. We can easily change this to return a value like $XXX,XXX instead of a null value, by changing the value of that case statement.

```
CREATE SECURE VIEW PROD.EMPLOYEES_CLEANED.SALARIES AS (
SELECT
EMPLOYEE_NAME,
JOB_TITLE,
CASE WHEN CURRENT_ROLE() = 'HR_ADMIN' THEN SALARY
ELSE '$XXX,XXX'
END AS SALARY_MASKED
FROM PROD.EMPLOYEES.SALARIES
);
```

This isn't particularly exciting, but it shows that it is just as easy to manipulate that return value as it was when creating masking policies.

The next example we're going to walk through is for the table PROD.EMPLOYEES.CONTACTS with the columns employee_name, email, and phone. We want to mask the email so that users without the role HR_ADMIN see only the domain, and we want to mask the phone number so that those users see only the area code.

```
CREATE SECURE VIEW PROD.EMPLOYEES_CLEANED.CONTACTS AS (
SELECT
EMPLOYEE_NAME,
CASE WHEN CURRENT_ROLE() = 'HR_ADMIN' THEN EMAIL
ELSE REGEXP_REPLACE(EMAIL, '.+\@', '*****@')
END AS EMAIL,
CASE WHEN CURRENT_ROLE() = 'HR_ADMIN' THEN PHONE
ELSE SUBSTR(PHONE, 0, 5) || '***-****'
END AS PHONE
FROM PROD.EMPLOYEES.CONTACTS
);
```

We can easily mask multiple columns at a time. We can also easily combine row-level logic at the same time, by including these statements in the where clause. Since these are evaluated at the same time, we don't need to worry about masked values within this view interfering with row-level access control.

Creating a Python Script to Generate Secure Views

We can create a Python script to generate these views more easily, if we think this is something we will spend a bit of time doing. Since we need to selectively mask each column in the SELECT clause, we either need to store the configuration for each table in code, or we need to pull that data from Snowflake. We're going to assume we have a function that returns the columns in a table in our Snowflake instance so that we don't need to recreate that for the purpose of our script.

First thing we're going to do is to define the parameterized SQL. We only need one SQL query to create the view.

```
SECURE_VIEW_SQL = """
CREATE SECURE VIEW {database}.{schema}.{view} AS
(SELECT
```

```
  {columns}
 FROM {table}
);
"""
```

Now we can define the inputs for our function and include a description of how the arguments should be formatted.

```
def create_secure_view(
    database,
    schema,
    view,
    table_name,
    column_role_mappings,
    ):
    """
    :param database: the database for the view
    :type database: str
    :param schema: the schema for the view
    :type schema: str
    :param view: the name of the view
    :type view: str
    :param table_name: the fully qualified table name
    :type table_name: str
    :param column_role_mappings: a column and its role to output
    mapping i.e.,
        {'email': {'role_maps': [{'role': 'HR_ADMIN', 'return_val':
        'email'}], 'else': 'NULL'}}
    :type logical_statements: dict
    """
```

We could just as easily split database and schema into two separate parameters, so we could specify a source database and schema, and a target database and schema rather than assuming that they will be the same. Next, we need to do the transformation step. This is a little trickier than with row-level access since we will need to specify every single column and only substitute logic for the ones that need masking.

```
cols = get_table_columns(database, schema, table_name)
select_clause = []
for col in cols:
    if column_role_mappings.get(col):
        col_logic = ['CASE ']
        mappings = column_role_mappings.get(col).get('role_maps')
        for mapping in mappings:
            col_logic.append(
                'WHEN CURRENT_ROLE() = {r} THEN {v}'.format(
                    r=mapping['role'],
                    v=mapping['return_val'],))
        col_logic.append('ELSE {v}'.format(
            v=column_role_mappings.get(col).get('else', 'NULL')))
        col_logic.append('END AS {c}'.format(c=col))
        select_clause.append('\n'.join(col_logic))
    else:
        select_clause.append('col')
select_clause_sql = ',\n'.join(select_clause)
```

Finally, we will return the SQL we generated so that it can be run on Snowflake.

```
return secure_view_sql
```

We could also combine this with our script for creating secure views with row-level access control, so that we could optionally do both within the same view.

Using Tokenization

Using external tokenization is significantly more complicated than using masking policies alone or using secure views. External tokenization also requires more moving pieces. To use external tokenization to mask column data, we first load tokenized data into Snowflake. We can then create the infrastructure to support Snowflake external functions in Amazon Web Services (AWS), Azure, or Google Cloud Platform (GCP). Once we create the infrastructure, we can create external functions in Snowflake that reference the work we did in our cloud provider. These external functions will be used to detokenize values so that certain roles can access the original source data. The final step is to create a masking policy that actually de-masks the data, and then applying that on the columns we've tokenized.

When we pre-tokenize data before loading into Snowflake, we can use any method of tokenizing that we prefer, as long as it matches our use cases. If we don't have a need to detokenize the data, then we can use any hashing function on the data, as long as it is safe and reasonably difficult or impossible to undo. Since we won't be detokenizing, we don't need a function that can be reversed. If we want the data to be recoverable, then we need to use a function that can be reversed. There are many libraries that exist for this purpose, so I recommend using one of those rather than implementing from scratch. I prefer to use Python's hashlib as it is very user-friendly and has most of the functions I might want to use. Once we've run all of our data through this function, we can load it into our Snowflake table. Since the data is already tokenized, it is safe, though it is not possible to detokenize and view the original data yet.

Once we have our tokenized data in a Snowflake table, we can start working with our cloud provider. Right now, Snowflake supports external functions using AWS, Azure, and GCP. We are not going to go very in-depth into how to create the infrastructure on the cloud provider side since that is a bit out of the scope of this book and may change. Regardless of cloud provider, the process is similar. We essentially need to create an endpoint that is configured for access from Snowflake, that calls a function that is run externally on our cloud provider. We will also need to create an API security integration in Snowflake. The syntax of this differs depending on the cloud provider. We can then use that endpoint and integration to create an external function in Snowflake. The syntax is the same for all of the providers.

```
CREATE OR REPLACE EXTERNAL FUNCTION <FUNCTION NAME>(PARAMETER_1, ...)
    RETURNS <RETURN TYPE>
    API_INTEGRATION = <INTEGRATION NAME>
    AS '<FUNCTION URL>';
```

Now that we have the function created, we can invoke it. Since this will be used on a column-by-column basis to demask certain columns, we want to call the function within a masking policy. Just as we used logic to determine what manipulation we will do on a value in the masking policies earlier in this chapter, we will create a masking policy that uses logic to determine when to call these external functions. We're going to create an external function called DETOKEN_SALARY that returns detokenized salary data and uses an API integration we can assume we already have set up.

```
CREATE OR REPLACE EXTERNAL FUNCTION DETOKEN_SALARY(SALARY)
    RETURNS STRING
    API_INTEGRATION = CLOUD_API_INTEGRATION
    AS 'https://cloud_platform-us-east-1.com/detoken_salary';
```

Our demasking policy for salary data will look like

```
CREATE MASKING POLICY DEMASK_SALARIES AS (SALARY VARCHAR)
RETURNS VARCHAR ->
CASE
WHEN CURRENT_ROLE() = 'HR_ADMIN' THEN DETOKEN_SALARY(SALARY)
ELSE '$XXX,XXX'
END;
```

Now that our policy is created, we can apply it to tables and views in the exact same way that we did with dynamic masking policies earlier. With external tokenizing, most of the work is done outside of Snowflake, which may mean that a different team will work on the cloud provider component than the team creating the masking policies in Snowflake. Your security or infrastructure team will be the experts on how to best set up these integrations so that they work with how everything else is set up in your organization.

Caution   When creating an external function that unmasks data, make sure that the function is not available to use by anyone other than the role that creates the demasking policy. Otherwise, the function could be used on its own to detokenize the data.

Combining Column-Level and Row-Level Permissions

We can combine row access policies and column masking policies on the same Snowflake object. This allows us to filter both the rows that a user will see as well as the fields they will see. We can solve problems like allowing managers to see certain employee data for their direct reports, without being able to see certain fields like SSN.

It is important to note that we need to be thoughtful about how we combine these policies. The row access policy is evaluated first, and then the masking policy or policies are evaluated. The same column cannot be used in both a row access policy as well as a masking policy, and the most recent of these added to the same column will create an error in Snowflake. As a result, it makes sense to use as few columns as necessary for a row access policy, and to avoid referencing any sensitive columns. It's also important to remember that this includes applying masking policies to tables referenced as part of another policy.

Tips and Tricks

Just like with row access policies, there are some tips and tricks we can use to make managing column-level access control easier. Some of the same tips and tricks apply from row-level access as well.

Categorize Data Types

To make managing column-level access control easier, I recommend categorizing the different types of sensitive data, if that is not already done for you. This means coming up with a list of types of sensitive data that should be externally tokenized, if that is something your organization will use, a separate list of data that should be masked but that is safe to be stored without tokenization, and data that can be optionally masked but doesn't necessarily need to be. This way we can decide what data is protected by the data team, and which data can be optionally secured by analysts or business owners. You can also utilize tags, which help identify data that has been categorized.

Create UDFs

In addition to the UDF we outlined in the last chapter, there are many other UDFs that we can create for column-level masking. I would recommend creating one or more UDFs that return partially masked data, if that is something your organization plans on doing. This way we can abstract away the transformation portion so that a mask just includes the logic for who should see what.

Use Consistent Naming Conventions

A mask or UDF name should be descriptive enough that a user knows exactly what it is doing without having to peek inside. This is especially important for masks since they are primarily used on very sensitive data and are often reused for multiple fields in multiple tables and views. If we're using external tokenization and have demasking functions, then it is important that those are clearly marked as doing the opposite of a mask so that they are not invoked when attempting to mask a column.

Key Takeaways

In this chapter, we saw the similarities and differences between implementing row-level access control and implementing column-level access control, and how we can effectively use both in tandem. Just like with row-level access control, we saw how powerful secure views can be in addition to masking policies. In the next chapter, we will explore how we can share data across multiple Snowflake accounts. The following are the key takeaways from this chapter:

- Column-level permissioning means replacing an entire column's data with null values, spoofed data, or tokenized data in a query's results.

- Dynamic Masking Policies are policies that sit directly on top of tables and views to protect the source data.

- Secure views can mask data by utilizing case statements in the select clause; however, the source data must also be secured.

- External tokenization is the safest way to mask sensitive data since it stores tokenized data, which is de-tokenized for certain users.

- External tokenization requires a lot of work with a cloud provider and involves creating masking policies to detokenize data.

- Row access policies are evaluated first, then masking policies, and a column cannot be referenced in both.

PART IV

Operationally Managing Access Control

PART IV

Operationally Managing
Access Control

CHAPTER 15

Secure Data Sharing

In the last chapters, we've covered how to create permissioning at the account, database, schema, table, row, and column levels. We've learned how we can structure our permissions so that the different levels work together, and how to best approach different problems. In this chapter, we're going to cover data sharing across multiple accounts, which we can see in Figure 15-1.

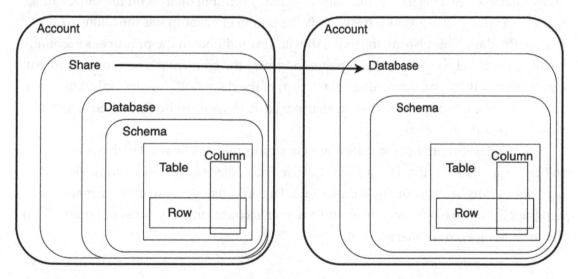

Figure 15-1. *Sharing data across accounts using a data share*

We will lean on the learnings from the past few chapters, since those privileges will need to extend to a data share that we will create.

What Is Secure Data Sharing?

Secure data sharing is a Snowflake feature that allows select datasets to be shared across multiple accounts. Data sharing is a read-only transaction between a producer account and one or more consumer accounts. Only certain account types can use

© Jessica Megan Larson 2022
J. M. Larson, *Snowflake Access Control*, https://doi.org/10.1007/978-1-4842-8038-6_15

secure data sharing, since it is considered a premium feature. We can also use secure data sharing with special reader accounts to allow read access to data without a high overhead.

Secure data sharing means sharing data across accounts owned by the same organization, or across multiple organizations. This can be used to provide a vendor or partner organization with information needed to be shared between the two, such as a product sharing their inventory with a grocer, or it could be used within an organization with many departments, each with their own instance. This could also be used to separate production and development, which we will cover in the next chapter.

Data sharing is a read-only transaction. With regular Snowflake tables, unlike external tables, data is stored on Snowflake-hosted storage. When data is shared across accounts, the consuming account will access the data on the producer's partition of Snowflake storage. This means that they're actually running queries on the same storage that the producer is using. As a result, only the producer account has the ability to mutate the data. This also means that if the data is modified on the producer's account, it will immediately be reflected when querying from the consumer's account, because it is referencing the same data rather than a copy of the data. Additionally, only tables and secure views are supported with data sharing; basic views must be converted to secure views to include in a share.

Snowflake also provides special reader accounts that can be used with data shares, so that organizations that do not primarily use Snowflake can use an account for read only operations like consuming a data share. The producer account creates, manages, and pays for these reader accounts, and a reader account can only consume shares from the producing account that created it.

Why Use Secure Data Shares?

Data shares allow accounts to safely share data across multiple accounts, including with external organizations. This is a much cleaner process than traditional ways of sharing data across organizations. Organizations may also choose to separate data across multiple accounts and wish to share some data across accounts.

When two organizations need to share data with each other in order to expedite their business, the typical way of doing this is quite involved. The process usually involves a data engineering team writing custom Python code to extract a dataset from the database, dumping the data to a file, and then dumping that file into the consuming organization's cloud storage, or worst case, sending it through an email. This is especially problematic because sensitive data should never be sent through email. This job would typically run on a daily or weekly basis, and might need to be modified as needs change, or if there are multiple customers requiring reports using different cloud providers, and therefore, more custom code. Using secure data sharing, the process is safer as it all takes place in Snowflake, and it requires less work on the part of engineers. Ultimately, this means less risk and less cost to both organizations.

As we will cover in the next chapter, this can also be used within an organization to create separation between different collections of data. This could mean a large organization or conglomerate with many different entities having a separate account for each entity that may need to share some subsets of data across entities. This could also mean having separate accounts for development and production. In either situation, data is clearly separated across accounts, and only the necessary data is shared.

How Do I Use Secure Data Shares?

Using data shares requires a bit of administrative work and careful consideration, as unintentionally sharing either the wrong data, or the right data with the wrong account, could be costly. Both the provider account and the consumer account will need to do work at their ends to complete the data sharing handshake.

Provider Account

The responsibility of the provider account is to create the share, allocate objects to it, and share it with the consuming account. At any point in time, the provider account can stop sharing any object with the consumer, can stop sharing the entire dataset with the consumer account, or can drop the share. Likewise, at any time the provider can add data to the share, extend the share to other consumer accounts, or create new shares.

Create a Share

In order to create a share, you must assume a role with the CREATE SHARE privilege on the account. You can create a share using the user interface or using SQL. I have a preference for SQL because it is consistent and repeatable, whereas a UI can change and the steps to create the share could be different next time. Since shares are account-level objects, we don't need to store them in any particular database or schema for shares, we can give them a simple but descriptive name. To create a share, run the following:

```
CREATE SHARE <SHARE NAME>;
```

This will create a share without access to any objects.

Grant Object Privileges to Share

Next, we must grant access to objects in Snowflake to our newly created share. We cannot add an account to a share until we have granted USAGE on a database to the share. Just like when we grant access to roles, we need to grant access from the top down. We'll grant USAGE on the database, USAGE on any schemas we may want to include data from, and then the privileges on tables and views.

```
GRANT USAGE ON <DATABASE NAME> TO SHARE <SHARE NAME>;
GRANT USAGE ON <SCHEMA NAME> TO SHARE <SHARE NAME>;
GRANT SELECT ON <TABLE NAME> TO SHARE <SHARE NAME>;
```

Now that we've allocated object privileges to the share, we can double check that we've correctly allocated access.

Verify Share Privileges

The first step to verifying the privileges granted to a share is to check the grants that exist.

```
SHOW GRANTS TO SHARE <SHARE NAME>;
```

This will give us the same information we may typically get when we query for the grants to a role. Specifically, we want to make sure that the privilege and the object match up, so we have USAGE on the correct databases and schemas and SELECT on the correct tables and views.

We can additionally verify that our secure views work correctly by simulating that we are using the share from a separate account. First, we set a parameter:

```
ALTER SESSION SET SIMULATED_DATA_SHARING_CONSUMER = <ACCOUNT NAME>;
```

We'll set the account name to be a name of an account we want to impersonate for the purpose of querying secure views, whether materialized or not. Currently using secure UDFs is not supported with impersonation.

Optionally Create a Reader Account

If we want to securely share Snowflake data with an organization that does not currently use Snowflake, we can create a reader account for them. A reader account is a special type of Snowflake account that is restricted to certain actions and is managed and paid for by the owning account. We can create a reader account using the user interface, or we can create it using SQL.

```
USE ROLE SYSADMIN;
CREATE MANAGED ACCOUNT <ACCOUNT ALIAS>
  ADMIN_NAME = <USERNAME>,
  ADMIN_PASSWORD = <SECURELY GENERATED PASSWORD>,
  TYPE = READER;
```

Once we create the account, Snowflake will return the account name you will use to reference the account when sharing data. This account will likely look like RE12345. We can run the following to see this information again:

```
SHOW READER ACCOUNTS;
```

The locator field will be the value needed to reference an account from a share. This reader account can only be used with shares from the parent account.

Share with External Account

Now that we've created the share, granted object privileges, verified the privileges, and created our reader account if we needed to, we can add accounts to our share. To add an account, we can run

```
ALTER SHARE <SHARE NAME> ADD ACCOUNTS=<ACCOUNT NAME>;
```

The command is the same whether it is a fully fledged account, or simply a reader account. We want to be careful that when we want to add an account to a share, we use ADD rather than SET because SET acts to reset the accounts on a share and will remove any existing accounts not in the statement.

Consumer Account

Now that the producer has created and shared the secure data share, we need to do some work at the consumer end to finish pulling the data in. We will identify the share as an inbound share, create a database from it, and then grant privileges to our users so that we can take advantage of the data shared with us.

Create Database from Share

The first thing we will do to create a database from the share, is to execute the following:

```
SHOW SHARES;
```

All shares that have been created in this account or shared with this account will show up in the results. Shares created by our account will have OUTBOUND in the kind column. Shares that have been shared with our account will have INBOUND in the kind column. If the database_name is null and the kind is INBOUND, then we know that the share has not been used in a database yet, and we can create one from it. To do that we can run the following:

```
CREATE DATABASE <DATABASE NAME> FROM SHARE <PRODUCER ACCOUNT>.<SHARE NAME>;
```

The <PRODUCER ACCOUNT>.<SHARE NAME> should match the output in the name column returned by SHOW SHARES. Now we have a database created from our share, and we are almost ready to have our users interact with the newly shared data.

Grant Privileges on Share

A role that either owns the newly created database from the share or a role with the MANAGE GRANTS privilege can grant privileges to users in the consuming account. To do that, we can run the following:

```
GRANT IMPORTED PRIVILEGES ON DATABASE <DATABASE NAME> TO ROLE <ROLE NAME>;
```

Since databases created from shares are different from typical Snowflake databases, we manage the grants through IMPORTED PRIVILEGES rather than a single privilege at a time. This means that all of our users will have the same privileges on the share. If we desire more precise privileges than this, we can grant the IMPORTED PRIVILEGES to a smaller group of users, and then create secure views from this data, just as we did with row-level and column-level access control.

Revoking Access and Dropping Shares

When we decide to revoke objects from a share, accounts from a share, or drop the share entirely, the effects are applied immediately. Since the data isn't copied to the consumer accounts, the data is instantly made unavailable. Additionally, since we grant privileges to a share, and that share is then granted to an external account, we cannot grant privileges to particular accounts, all accounts have the same access unless we supplement with secure views and logic that explicitly invokes the account name. If we want to revoke access to a subset of data in the share from a particular account, we should revoke access to the share, and create a new share with that subset of data.

Revoking Object Privileges from a Share

When we revoke privileges on an object from a share, we can run a revoke command in the same way we would with any other type of object from a role. To revoke select on a table from a share, we can run the following:

```
REVOKE SELECT ON <TABLE NAME> FROM SHARE <SHARE NAME>;
```

Any account with access to this share no longer has access to this particular table.

Revoking Access to a Share

To revoke access to a share from an account, we can run an ALTER SHARE command just like we did to add the account.

```
ALTER SHARE <SHARE NAME> REMOVE ACCOUNTS=<ACCOUNT NAME>;
```

If we aren't sure which accounts have access to the share, we can first run the following:

```
SHOW GRANTS OF SHARE <SHARE NAME>;
```

The name of the account to use in the ALTER SHARE command will show up in the account column.

Dropping a Share

We can drop a share using the DROP SHARE command. This does not require removing access to the share or removing objects from the share. To drop a share entirely, removing access for all accounts instantly, we can run the following command:

```
DROP SHARE <SHARE NAME>;
```

Effective immediately, consuming accounts will no longer be able to query data from this share.

Constraints

Using secure data sharing seems pretty straightforward, but there are a few constraints we need to be aware of. Since secure data sharing takes place across separate accounts, we cannot guarantee that anything will be the same between the accounts. As a result, we should be aware of how those differences affect data sharing.

Secure Views

Only secure views are allowed to be included as part of a share, basic views are not permitted. Additionally, every object referenced within a secure view must exist in a database included in the share. If a table from a separate Snowflake database is referenced in the view, any queries against it will fail. For this reason, I recommend storing the raw data in schemas that are not shared with the other account and pulling data from those schemas. We're going to create our view as follows:

```
CREATE SECURE VIEW PROD.CLEANED_VIEWS.TEST_SECURE_VIEW AS (
  SELECT COLUMN_ONE,
  ...
  FROM PROD.RAW.TEST_TABLE
);
```

We can then grant USAGE on the database and the secure view's schema to the share, omitting USAGE on our private schema, RAW from the share.

```
GRANT USAGE ON DATABASE PROD TO SHARE <SHARE NAME>;
GRANT USAGE ON SCHEMA PROD.CLEANED_VIEWS TO SHARE <SHARE NAME>;
```

```
GRANT SELECT ON PROD.CLEANED_VIEWS.TEST_SECURE_VIEW TO SHARE <SHARE NAME>;
```

This allows us to protect the table while still conforming to the secure data sharing requirements.

Cross-Region and Cross-Cloud Storage Platform

When sharing data with an account that uses a different region, or an account that uses a different cloud storage platform than the producer account, external tables are not supported. If one or more external tables exist in a database that is included in a share, accounts with a different region or cloud platform will not be able to create a database using the share. This includes tables that a share does not have privileges on. To prevent this from happening, I recommend avoiding using external tables with databases that will be included in a data share, if possible.

CURRENT_USER and CURRENT_ROLE Functions

When we create secure views and UDFs that will be used in a share, we need to remember that the logic will be invoked in an entirely separate environment. Usernames will probably not have the same format, and depend on the context of the account, and so using the CURRENT_USER function is not allowed. Additionally, we cannot guarantee that the consuming account will use the same roles as we have set up in our primary account. Therefore, we should not use the CURRENT_ROLE function either.

Granular Access Control with Shares

Since we have constraints with more granular access control when we work with shares that we don't have when we manage permissions within a single account, we need to use different strategies to accomplish access control. Before we get started, I want to stress that we should assume that any data we include with a share could be accessed by anyone in any of the organizations we have included as part of the share. We can rely on certain Snowflake functions to determine which account is querying the data, which will allow us to separate data by organization; however, we do not have any control around what happens within an account.

Access Control by Account

We can broadly control which accounts can access subsets of data using secure views. To do this, we want to rely on the built-in Snowflake function CURRENT_ACCOUNT. We can create a secure view with row-level access control using this function like

```
CREATE SECURE VIEW <VIEW NAME> AS(
  SELECT <COLUMN>,
  ...
  FROM <TABLE_NAME>
  WHERE ACCOUNT_NAME = CURRENT_ACCOUNT()
);
```

Since this is a secure view using a built-in function, we can trust that our data will be safe. We can also use the INVOKER_SHARE function to return the name of the share being used.

```
CREATE SECURE VIEW <VIEW NAME> AS(
  SELECT <COLUMN>,
  ...
  FROM <TABLE_NAME>
  WHERE <SHARE 1> = INVOKER_SHARE()
  OR <SHARE 2> = INVOKER_SHARE()
);
```

This is particularly useful if we include a secure view in more than one share.

Consuming Account RBAC

If the consuming account wants to restrict usage of the data in the share from users within their Snowflake account, they can create secure views downstream of the original data. They should then limit the users that are granted IMPORTED PRIVILEGES on the database, and instead, grant users USAGE and SELECT on secure views placed in a different database and schema in their account.

Key Takeaways

Data sharing is a unique feature to Snowflake that can allow large organizations, or organizations that work with many other entities to share data in a secure fashion, while not sacrificing data quality, data availability, or data latency. In the next chapter, we will dig into how we can separate production and development environments and will spend a section on how we can use data sharing to accomplish this. The key takeaways from this chapter are the following:

- Data sharing involves a producer account – the account that is sharing their data – and a consumer account – the account receiving the data.

- The consuming account is allowed read-only access to tables and secure views included in the data share.

- We can create reader accounts for external organizations that may need to use our data but do not have Snowflake accounts.

- Data sharing reads all data from the Snowflake internal storage of the producer account.

- External tables cannot exist in a database that is shared with an account using a different region or cloud storage platform.

CHAPTER 16

Separating Production from Development

In the last few chapters, we covered the different levels of access control. Now that we understand how we can handle permissions on the account, database, schema, table, row, and finally column level, in addition to working with data shares, we can touch on separating development from production. We will lean on concepts we learned in these chapters in order to craft separation between development and production environments.

What Does It Mean to Separate Production from Development?

When we separate production from development, a process commonly referred to as prod/dev separation, we're identifying one environment as the primary one used to power reporting and other analytics use cases at the company, and one to be used to develop and test planned changes to the system. We can separate production from development by using an entirely separate database and account instance, or we can separate it using different databases within the same Snowflake account. We can copy production data into the development environment, or we can include fake data that looks and feels like the actual data. We can also structure environments so that read privileges in production are mimicked in development.

When we initially set up our separate environments, we can decide just how separate we want them to be. It is possible that just using separate databases within our production system is sufficient. Certain organizations may prefer to have an entirely separate account for development. It may be the case that your organization uses a combination of these methods to support different types of development, by allowing

© Jessica Megan Larson 2022
J. M. Larson, *Snowflake Access Control*, https://doi.org/10.1007/978-1-4842-8038-6_16

a separate account to be a development instance for the platform team, and a separate dev database in the same account as production for other use cases, like transformation. A very simple setup may mean separate schemas that teams use as sandboxes for their development data.

When we populate a development environment with data, we can either populate it with data from production, or with synthetic data. If we populate dev with real data, we need to keep in mind the privileges on the data so that we do not create an access loophole where unmasked and unfiltered data is available in dev to users that would not otherwise have access in production. For this reason, it is becoming increasingly common to see spoofed or synthetic data. This means using a tool or service to generate fake data that looks and feels like the source data. This allows users to play around with data during development without exposing sensitive data.

In the case that we populate dev with production data, we need to make sure that we have the same or similar privileges set up on both ends. This is mainly important for read access, we may be more generous with write access on the dev environment since the risk of affecting production does not exist. We also want to enable our developers working with data at all stages to work efficiently and feel empowered to do their best work, rather than having them feel like they need to jump over hurdles to do their job. For this reason, we want the privileges on dev to be seamless, pulling configuration from the production instance.

Why Separate Production from Development?

There are many reasons why prod/dev separation is a topic engineering leadership loves to talk about. When we maintain separate environments for production and for development, we can better ensure the quality of data powering important business decisions, which in turn means less downtime and brings our organization greater value – helping us gain and retain trust in data. Regulations like SOX also dictate that we may need separate environments to complete an audit of changes before they take effect. Separating environments can also help our developers move quickly as they will not need to worry about the sanctity of the production environment; when developing in a closed development environment, they can move fast without breaking things.

When we maintain a separate development environment, we can make sure that none of the data powering important business decisions is modified or corrupted. This means developers can modify the definition of tables or views that provide data for reporting, without actually affecting the reports since the changes are made on a copy of those tables and views. This means more uptime – no scrambling to fix a report that was inadvertently altered. This also means we can more easily build trust with our stakeholders by minimizing data quality issues and improving the quality of service. We can even set up a development instance of our downstream tools we may have for things like reporting and visualization, then compare the newly updated reports to those that exist in production to verify that the resulting data is the same.

Regulations like SOX may mean that your organization must separate development from production to meet requirements. SOX internal controls often dictate a process that must be followed to audit and validate that the changes to underlying data do not change the results seen in a dashboard or report. This requires making changes within a development environment, duplicating the other settings within a report, and validating that the results are the same as we see in production, or the changes are what are expected.

When we separate dev and prod, we're supporting our developers. Making changes to data in production requires a lot of caution, checks, and care, which in turn means more time and attention from developers. By providing an environment as similar to production as possible, but with guard rails, we allow our developers to focus on what matters most – the problem they're trying to solve with data. Since the environment is entirely separate, they will not need to worry about contaminating production data, and won't need to worry about accidentally dropping tables, views, or schemas.

How Do I Separate Production from Development?

When we implement prod/dev separation, we first start by identifying the problem we're trying to solve and outlining requirements for the use cases we're trying to support. Once we've done that, we can create a dev environment that works for our problem space. Then we can start to populate data or come up with a strategy for populating data. Before users start to access development datasets, we need to make sure that our access control privileges are sufficient. Finally, we can start to integrate our development environment with our existing tools and services, preferably development environments within these tools and services.

Types of Users to Support

When we create a development environment, we have a few different types of users that we will need to support. In some organizations, these will be combined into fewer than four distinct use cases. What we're trying to assess here is who we need to support, and how we can best support them.

Platform Developers

Platform developers are the ones primarily managing the Snowflake instance, which, if you're reading this book, is very likely your team. We're the users who test out newly released Snowflake features, or even those still in beta. Since these features may change the behavior of the entire rest of the database, it may make sense for these users to have a dedicated instance. Since we want production to be as similar to development as possible, having these users share a dev instance with other developers might interfere with and invalidate their testing.

Pipeline Developers Working with External Sources

Pipeline developers work with tools and APIs to pull data from an external source into Snowflake. When they're developing and testing a new pipeline, they may need to run the process multiple times loading data into one or more tables. When they're making changes to an existing pipeline, it may mean changing the table's definition, and doing a backfill for a new column. Both of these workflows can be done in a development environment using production data.

SQL Developers Transforming Data

SQL developers are creating downstream tables and views from data already in Snowflake. They may be working on data modeling, or simply creating some pre-joined tables for use in dashboarding and reporting. These users may need access to the original source data in order to get their work done in some cases, but may be able to use spoofed data in many cases. These developers are our primary users we are trying to support in the next few sections.

Visualization Developers

Visualization developers work on the final leg of the data process. They're creating visualizations and reports in a reporting tool like Tableau or Looker. For performance reasons, these developers may create downstream tables and views like the SQL developers, but will additionally need to pull this information into a report using the preferred tool of the organization. To support these users, we need to make sure that we can connect our development infrastructure to either the production version or a separate development instance of this visualization tool.

Creating a Dev Environment

When we create our developer environment, we need to keep in mind how we want to approach prod/dev separation. We can think of prod/dev separation as a spectrum ranging from not at all separate to fully separate. The smallest amount of separation we can have between development and production is by utilizing sandbox schemas for development. From there, we can move to using a different database within the same Snowflake account, which is more separation, but would not be considered fully separate. Then, to fully achieve prod/dev separation, we can use separate accounts. This section will be mainly geared towards our SQL developers who will need to work with copies of production data, or synthetic data that behaves similarly.

Using Separate Schemas

Using schemas to separate production and development is not considered prod/dev separation; however, it can work as a starting point, especially for organizations with small teams that may not have a need for full prod/dev separation. We can see how we can use separate schemas for production and others for development in Figure 16-1.

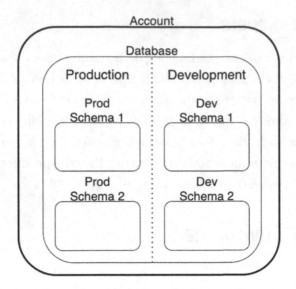

Figure 16-1. *Using separate schemas for development and production*

As we can see, using this system, we end up with multiple production schemas and multiple development schemas, which makes the separation a little less clear. I want to be clear that this is not a best practice; however, it may be something that makes sense for your organization at its current size. The nice thing about using separate schemas is that it is really easy to set up. Once we decide on a naming convention to differentiate the development schemas from the production schemas, we just run

```
CREATE SCHEMA <SCHEMA_NAME>;
```

It is important that we choose a naming convention that makes it very clear to everyone that the data is not validated and should not be used in any reporting or dashboarding. I recommend prefixing DEV or similar.

Using Separate Databases

Using databases to separate development from production allows us to essentially create a separate namespace for development. Again, this is not a fully separate environment; however, it seems to be a great compromise between a fully separate account and separate schemas. We can see how we can use separate databases in Figure 16-2.

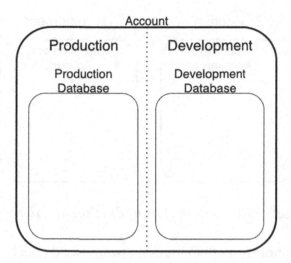

Figure 16-2. *Using Separate databases for production and development*

Unlike with separate prod and dev schemas, using separate databases creates a clear delineation between the two environments.

Snowflake allows cross-database joins, so it is important to note that this is not fully separate, and like with separate schemas, it is important that we choose a naming convention that makes it clear that the data in that database is not to be used for any reporting or dashboarding. If we want to work with production data, we will actually create this database at the same time as we clone the data, which we will cover in the next section. Otherwise, if we wish to use synthetic data, we can create a database like

```
CREATE DATABASE <DATABASE NAME>;
```

I recommend calling this database DEV or something similar, so it is clear to users that this database is not for production use.

Using Separate Accounts

Using separate accounts is the only true option for prod/dev separation. This creates two entirely separate environments that do not share any dependencies or parameters, as shown in Figure 16-3.

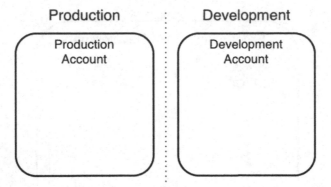

Figure 16-3. *Using separate accounts for production and development*

This requires a bit more work than the other two, as we do not have the ability to create Snowflake accounts on our own. To create a development account, you will need to reach out to your Snowflake representatives and ask them to create one for you. To take advantage of Snowflake's Data Sharing feature, I recommend creating this new account in the same region as your primary account. You'll want to configure everything else in the account to be identical to your production account as well.

Populating Dev Data

Now that we've created our development environment, we need to populate it with data. We can either populate it with real production data and make sure that we include all the same protections on the data or we can instead populate dev with spoofed data that behaves like the original data.

Production Data in Separate Schemas

When we use separate schemas for development, we may not need to copy data in order to accomplish our tasks, and instead, we just create copies within staging or development schemas. If we want to make changes to a dataset using copied data, we can either clone entire databases, or clone individual tables. To clone a schema, we can run the following, which will recursively copy all objects in the original schema into a newly created schema:

```
CREATE SCHEMA DEV_<SCHEMA NAME> CLONE <SCHEMA NAME>;
```

We can also copy an individual object like

```
CREATE <DATABASE NAME>.<SCHEMA NAME>.DEV_<TABLE NAME> CLONE <TABLE NAME>;
```

Since a table is the smallest container or objects, this will only clone the table.

Production Data in Separate Databases

To create a development database, we will start by cloning the production database. To do this, we run the following:

```
CREATE DATABASE DEV CLONE PROD;
```

In Snowflake, cloning is a zero-copy operation. The newly created clone of the primary database is also entirely separate from the primary database, meaning that any changes to either database will not be reflected in the other. Once an object within the clone database is changed, additional storage for that object will be created; before that happens, no additional storage is used by cloning. Additionally, cloning will not copy over external tables or internal stages.

When we develop with a dev database, we want to avoid using fully qualified object names so that we can parameterize the database in our scripts. This means our Python scripts and other tooling should detect whether the process is running in production, or whether it is running in development. We can either use templating to inject the development database into our scripts, or we can specify the database in our connection parameters.

Since cloning is a zero-copy operation, creating a development database is extremely cheap. Some organizations may find that instead of creating a single development database, they would rather create development databases for each user. This could be prompted using an internal tool where a user clicks a button, it could be done automatically for a certain group of users, or users could do this themselves.

Production Data in a Separate Account

Copying over data to a separate account for development purposes is a bit trickier than when developing in the same account. We essentially have two options: create a scripted process that copies over objects and their data or use secure data sharing.

Creating a scripted process to copy objects to the development account could be done in any number of ways. We could simply bake this into how we load data into our production database. This could mean prompting loads from the same cloud storage location in both accounts using bulk loading or through using snowpipes to automatically load data once it lands in storage. This is probably the easiest option for actually copying the data, though it will require an initial backfill. Another option is to have this on demand, where a prompt kicks off the process to load data from a cloud storage location into the development account. The benefit to doing it this way is that the data would be fully fledged data that can be modified in any way the organization wishes. However, this means that additional costs will be incurred for both storage as well as the loading and development compute.

The other option is through using secure data sharing, which we covered in the last chapter. Refer to last chapter for the steps to create and share data between accounts. Once we've created a database using the cloned data, we can start to create downstream objects from the source data. Since data sharing is read-only, we cannot modify existing objects, which may limit our development. We can, however, create copies of the data and make our modifications there. To do that, we can either use views, or create tables using the source data.

Synthetic Data

Instead of populating our development environment with production data, we can instead populate it with synthetic data. First, we classify the data we have in Snowflake, so we can identify which classes of data we need to generate for which columns. These classes would be categories like email, phone number, names, addresses. We can then populate the data ourselves using the Python connector, generating data and inserting it into our development environment. We could optionally use tooling to accomplish this process for us.

To classify our data so that we can insert synthetic data that matches the classes of data we have, we have a couple of options. We can classify manually, identifying the types of data and tagging accordingly. Snowflake also provides data classification, which identifies PII with some degree of accuracy. While this data classification tool is very useful, it may not catch every instance of PII and therefore should not be used on its own. There are also third-party tools that can scan tables and automatically tag columns with their types. We might also decide that we don't need to be so specific about the class of data, and that randomly generating strings or numbers would be sufficient. It is important that we identify which columns include identifiers, used as primary or foreign keys, that map to other columns so that we do not break those connections.

When we generate synthetic data in Python, we have a couple of options of how we can do this. We can use libraries like *pydbgen* to generate data that fits different classes of data. To do this, we simply create pandas dataframes using a data generating library with the data we want to see, and then load that resulting data into Snowflake tables. We could also generate random strings, numbers, and dates, matching the datatype of a column to the type of random information we insert.

There are also tools that are plug-and-play that will handle this process for us. Tools like *tonic.ai* will connect to a read-only copy of your production database and automatically load synthetic data into the destination database of your choice. Tools like this make it easier for smaller teams to provide prod/dev separation.

RBAC in Dev

If we're using production data to power our development environment instead of synthetic data, we need to be cautious that we did not inadvertently create a data leak. Production data should have the same read restrictions in the development environment as it does in the production environment. Since this environment will be used for development, we will make sure that users have write privileges on data they're allowed to change, but users should not be able to see any data in development that they wouldn't otherwise be able to view in production.

For All Setups

Whether we use a separate schema, database, or account, there are shared RBAC concerns we share. No matter how we set up our production and development environments, we want to make sure that users cannot mutate data in production, and likewise, we want to make sure that our production services do not have the ability to change data in development. We also need to make sure that if we are cloning data, the role cloning the data has the correct privileges to do so.

I recommend using a separate service account for cloning data from production to development than is used for ETL. This is to provide extra guardrails preventing errors in our ETL processes. Ideally, our production service account should only have write access on production, and no privileges on development. We can then create a service account specifically for development that has no privileges on production. This prevents workflows from accessing or creating data in development and mixing stale or corrupted data from development with production data.

The role cloning data from one container to another must have usage on the container and any sub-containers, as well as select on all tables and views that will be cloned. This is tricky because this level of access is a lot for one particular role to have. For that reason, it may be easier to have multiple development environments, so no single role or account has access to all of the data in Snowflake.

Separate Schemas

If we're using separate schemas as staging or development environments, we need to make sure that users have read access on the production schemas and write access on the development schemas. We may want to designate certain schemas for certain teams, where members of the team have write privileges, and non-team members do not have any privileges at all. Blocking external teams from viewing the data in development schemas prevents inadvertent mixing of production and non-production data.

Separate Database

When we use a separate database within the same account for development, the easiest way to copy data is through cloning the production database. Since cloning databases and schemas is recursive, the underlying object permissions are copied as well. However, we still need to grant privileges on the cloned database. We will also need to add privileges to the objects within the database so that users can mutate the data in the development environment.

Once we clone the database, we need to grant privileges on the new database. Ideally, we would have a single role for either all users or the subset of users that develop data assets. To do this, we simply grant usage.

```
GRANT USAGE ON DATABASE DEV TO ROLE DEVELOPERS;
```

Now our developers can interact with the DEV database in the same way that they can in production. Our problem still isn't solved because we want users to be able to mutate data in development that they are not able to mutate in production. What we essentially want is that if a user has read privileges on an object in production, they should have write access on that in development. There are a couple of ways we can accomplish this. We can either iterate through the privileges and, for each object or schema, we can grant write to the roles that have read or we can create some automatic mapping where we grant write access on development to the read roles from production. An example is shown in Figure 16-4.

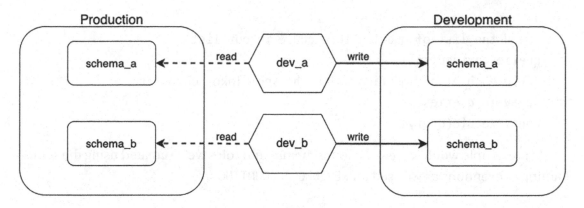

Figure 16-4. *Different developer roles in production compared to development*

In Figure 16-4, we can see that dev_a has read privileges on schema_a in production, and write privileges on schema_a in development, and dev_b has similar privileges on schema_b. In both cases, the roles still do not have any privileges on the other schema.

The cleanest way of handling this is to have a mapping between schema names and roles so that we can programmatically manage our grants. I recommend maintaining roles for schemas where each schema has its respective role, for example, if a schema were REVENUE_ACCOUNTING, it would have roles REVENUE_ACCOUNTING_READ and REVENUE_ACCOUNTING_WRITE. After creating our dev database, REVENUE_ACCOUNTING_READ should have the same privileges in development as REVENUE_ACCOUNTING_WRITE has in production. Since we followed these naming standards, we can create a really easy Python script to do this, though we could just as easily do this as a procedure in Snowflake if we want everything to live in platform. This script assumes that we have a function that takes the name of the role and the name of the schema and generates all of our write privileges, and that we've already created our connection to Snowflake.

```
cursor.execute('SHOW SCHEMAS IN DATABASE DEV;')
schemas = cursor.fetchall()
# filter out the columns we're not interested in
schema_names = [x[1] for x in schemas]
queries = []
for s in schema_names:
    sql = generate_write_access(
        role=s+'_read',
        schema=s,
        is_prod=False,
```

```
    )
    # add the list of queries to execute to our list
    queries.extend(sql)
# execute each of our queries using the snowflake connector
for query in queries:
    conn.execute(query)
```

This example would cover all of our schemas and roles we've created using the same naming conventions as we used for REVENUE_ACCOUNTING.

Separate Account

The tough thing about using a separate account for development is that no matter how we copy our data over, we do not maintain any of our original RBAC. In any situation, we will essentially need to reconstruct our RBAC from scratch. This leaves us with a few main options: we can allocate access in our development account at the same time as we do on our main account, we can pull our grants using system tables Snowflake provides and parse then reformat the results, or we can log all of our queries we run for access control on production and run those same queries on our development account.

If we decide to keep our development account up to date with all of the changes we make in our production account, we can additionally update our RBAC model as we go, too. This means running all of the same queries and operations on both accounts. As with copying the data, this will likely require a backfill, which we can do the same as one of the following two methods.

We can mimic the same RBAC as we have on production in development by using Snowflake system tables or system queries to find out the privileges that exist, and then we can parse the results and format them into new queries. To find all of the queries that exist within a database, we can run

```
SHOW GRANTS ON DATABASE <DATABASE NAME>;
```

This will give us all of the grants on that database with the following fields:

- created_on – The time the grant was executed

- privilege – The name of the privilege granted

- granted_on – The type of object the privilege was granted on (i.e., database or schema)

- name – The name of the object the privilege was granted on

- granted_to – The type of entity the privilege was granted to (i.e., role or share)

- grantee_name – The name of the entity the privilege was granted to

- grant_option – Whether or not the privilege was granted with the grant option

This is enough information to reconstruct our privileges into SQL. We can simply iterate through all of it in a simple Python script like

```python
privs = conn.execute('SHOW GRANTS ON DATABASE PROD;')
privs_sql = []
grant_sql = """GRANT {privilege} ON {obj_type} {obj_name} TO
    {grantee_type} {grantee_name} {grant_option};"""
grant_option_sql = 'WITH GRANT OPTION'
for p in privs:
    sql = grant_sql.format(
        privilege=p[1],
        obj_type=p[2],
        obj_name=p[3],
        grantee_type=p[4],
        grantee_name=p[5],
        grant_option=grant_option_sql if p[6] else '',
    )
    grant_sql.append(sql)
for query in privs_sql:
    cursor.execute(query)
```

This script assumes that we already have a Snowflake connection. We could put this in a loop for each database that we would like to copy to development, or we could simply change the first line to be SHOW GRANTS ON ACCOUNT;. This will return a maximum of 10,000 records, which should be enough for most organizations, but if you have exactly 10,000 records returned by this query, you may want to look into either iterating through smaller objects or look into using the system tables in the Information Schema instead.

Connecting to Dev

Now that we've created a development environment for our developers, we need to make sure they have the ability to connect to the Snowflake development environment from their regular coding environment. For security reasons, I recommend using keypair authentication for users. This means uploading a private key to Snowflake and maintaining the key at a specified location on either the users' local machines, or their devapps – a dedicated cloud compute instance for testing. Once that is done, the infrastructure set up for creating workflows should have a switch that determines whether the workflow is running in development or production. If it's running in production, it should authenticate using a service account, and if its running in development, it should use the developer's private key to authenticate.

Another option is to have a service account created for each team, with the user and password stored in secure storage like *HashiCorp Vault*. Then, the same infrastructure that determines whether a workflow is run in development or production could use some logic to determine which keys to use for the current developer. One note of caution for using a setup like this is that it is possible that everyone on a team does not have the same read privileges on an account. It is possible that certain users may have data they cannot access through this method. It is also possible that this may expose data.

Key Takeaways

Separating production from development is an important step in maturing a data warehousing platform. In this chapter, we covered the different ways we can separate the two environments to fit any organization's analytics infrastructure. Environments can be split at the account, database, and schema levels using a few different methods. In the next chapter, we will cover strategies for working with upstream and downstream services. The following are the key takeaways from this chapter:

- Separating production from development means maintaining separate environments for developers and end users.

- Regulations may require your organization to have separate production and development environments.

- Separating production from development helps maintain data quality, uptime, and building trust within your organization.

- To fully separate production and development, we need to use a separate Snowflake account.

- We can approximate prod/dev separation by using separate schemas or databases within the same account.

- For developers using the Python connector, it is recommended that they connect via keypair authentication or password rather than a service account.

- Naming conventions can make extending write access in development easier because it can be done programmatically.

CHAPTER 17

Upstream and Downstream Services

We're getting to the last couple of chapters now that we've wrapped up access control within Snowflake. We've covered the different types of privileges that exist on the different object and container types within Snowflake. We discussed how to structure different types of roles, and how to grant them to users. We went through the ways to handle permissioning at the different levels in the object hierarchy. At this point, we've effectively covered all of access control inside Snowflake.

If Snowflake existed in a vacuum, we would be done, we could stop here and call it a day. But Snowflake does not exist in a vacuum. We need to poke holes into our perfectly controlled environment so that we can power business decisions. We need to insert data into our data warehouse, and we need to pull that data out in a way that respects all the work we've done so far.

In this chapter, we're going to zoom out so that we can look at our data analytics infrastructure holistically, as we can see in Figure 17-1.

© Jessica Megan Larson 2022
J. M. Larson, *Snowflake Access Control*, https://doi.org/10.1007/978-1-4842-8038-6_17

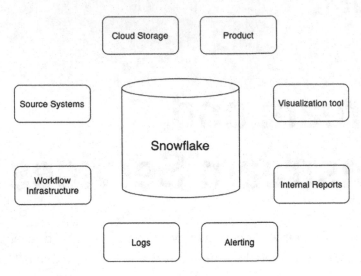

Figure 17-1. *An analytics environment with many different services surrounding Snowflake*

Upstream Services

Before our data gets to Snowflake, it's likely gone through a few different services or steps in a pipeline. Whether the original source of data is external – a ticketing or sales tool for example, or internal – data from the product, we need to ensure that the data is adequately protected on the whole path. We can visualize an example of typical upstream services in Figure 17-2.

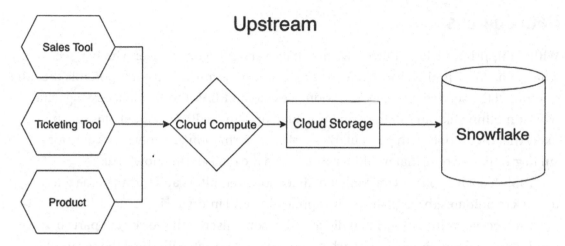

Figure 17-2. *Upstream services*

Figure 17-2 demonstrates a common pattern for upstream services for a data warehouse. We have a couple tools we're pulling data from, our organization's product we're pulling data from, the cloud compute we're using to orchestrate all of this, and the storage container we will use to hold the data to load into Snowflake.

What Are Upstream Services?

Upstream service is an umbrella term for any tool, system, or workflow data may pass through before it lands in Snowflake. Upstream services typically fit into a few different categories, data sources, compute and orchestration, and cloud storage. Each of these categories has one or more ways of enforcing privileges on Snowflake.

How Do I Maintain RBAC with Upstream Services?

Maintaining role-based access control in upstream services can be tricky, since each of these systems likely will have its own version of access control that may have different rules and mappings than exist within Snowflake. This makes it tricky to use Snowflake as the source of truth since these services handle data before data gets loaded into Snowflake. If we wanted to use Snowflake as the source of truth, we would need to pull configuration from Snowflake for reference in these systems, which is less desirable than maintaining these groupings elsewhere.

225

Data Sources

Within our primary data sources, we may have very little control over who has access to what data. We're limited by what features these services provide us. It is possible that one or more of the services your organization uses has an RBAC model for accessing data, with something like an API that can be used to update the RBAC model. It is much more likely that these tools with particularly sensitive information are only accessible for a small group of people that would have access to the datasets in Snowflake.

Tools like an applicant tracking tool that allows recruiters and hiring managers to track candidates through the entire application and interviewing process, often have an access control model that dictates that some users can see certain parts of the candidate's information based on what interaction they had with them. For example, a hiring manager might be able to see all of a candidate's information, but someone conducting the technical interview might only get to see a resume and their name. This would be an excellent opportunity to pull this information from the tracking software into a mapping table for row-level access control, so users can have access to the same data in Snowflake as they would if they used the original tool.

Compute and Orchestration

Compute and Orchestration is the worker and scheduler actually doing the work of connecting to the data sources from last section, potentially transforming the data, possibly dumping data to a cloud storage bucket, and then loading into Snowflake. This means that the worker will need to connect to Snowflake to prompt the load if it is not done automatically using a Snowpipe.

There are three main ways to authenticate from the worker to Snowflake: using a service account with a username and password, using a service account with keypair authentication, or using an OAuth token. Any of those will work great, as long as we securely store these keys, passwords, and tokens. Using a username and password is often the most convenient because these can both be stored in a keystore like *HashiCorp Vault* or an equivalent tool. Usually, these tools allow certain users or machines to unlock them. This allows the key to be invoked automatically, given the environment the process is running in.

Additionally, we need to decide what privileges our worker has within Snowflake. We can either grant our service account a role with write privileges on all datasets, or we can have our service account assume multiple roles depending on which workflow is being triggered. Then workflow IDs, for example, a *dag_id* in *Apache Airflow*, could be mapped to a role for the service account to assume. We could use a service we build to do the mappings where we check whether or not a user has access to a particular role before we allow them to map a workflow to a role. We could also include this as part of a manual code review process if we think the volume of workflows would be sufficiently low.

Cloud Storage

Often after data is extracted from the data source, it is dumped to a file and placed in one or more cloud storage buckets. Typically, a persistent storage bucket – that is, a bucket that will hold the data as an inexpensive backup for a long period of time, and a staging bucket – a bucket used to load the data into Snowflake with a very short time to live (TTL) for files. The short TTL means that data will only exist in that location for a short duration – typically 24 hours or shorter. Staging buckets can simplify the ETL process for bulk scheduled jobs by providing a space for files to land and persist only until the next run of the scheduled job; this allows us to easily rerun the final load portion of the job in the event of failure without worrying about additional files getting loaded. A staging bucket may also have a wider audience of engineers that need access, by having a short TTL, we're reducing the amount of data accessible.

Ideally, no users should be able to read files from the persistent cloud storage bucket, nor the staging bucket. The staging bucket should be created as a security integration in Snowflake, with a stage created from it. The security integration should be owned by a security admin or similar role, and only roles that will need to create stages should have USAGE on the security integration. To secure the storage bucket further, we can create our stages with a more specific path. For example, instead of creating a stage with access to the entire bucket, we can supply the parameter URL under `externalStageParameters`. Instead of a single `GENERAL_STAGE` with URL = `'<cloud provider>://<bucket name>'` we could instead create `BI_STAGE` with URL = `'<cloud provider>://<bucket name>/bi/'` and `HR_STAGE` with URL = `'<cloud provider>://<bucket name>/hr/'`. This prevents the `HR_STAGE` from pulling any data under `'<cloud provider>://<bucket name>/bi/'` and vice versa. Additionally, only service accounts should have USAGE on these stages in production.

Downstream Services

Once we have our data in Snowflake, we will need to connect to a service that allows us to visualize and work with our data. As we can see in Figure 17-3, there are a number of different tools and services we may have that pull data from Snowflake.

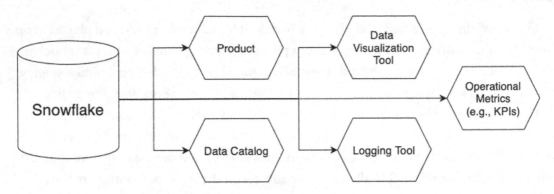

Figure 17-3. *Services downstream of Snowflake*

Since our downstream services pull data out of the Snowflake ecosystem into a separate environment, we need to take care that our downstream services cannot circumvent the protections we put in place.

What Are Downstream Services?

Downstream services are any tool, service, or system that works with data after it has been loaded into Snowflake. Examples of this include visualization tools like *Tableau* and *Looker*, data catalog tools like *Collibra*, and logging tools like *Splunk*. These are a wide range of tools that expose data in different ways. Some of these tools, like a data catalog, may not expose any data to users at all, instead, summarizing metadata. Other tools, like a visualization tool, are the main way that users interact with an analytics database, and therefore will expose any and potentially all data in Snowflake. It is important that we secure these tools in a way that does not expose Snowflake data to an unintended audience.

How Do I Maintain RBAC with Downstream Services?

Maintaining our role-based access control with downstream services can be very tricky because tools vary in their support of external or internal RBAC. Every tool is different, and certain tools lend themselves to RBAC more than others. Different tools also have different purposes and may only use metadata. When we think about downstream tools, we need to ask ourselves a few questions. What data will this tool expose to users? What data will this tool consume? Who will be using this tool? Does the tool support Snowflake's access control, or does it support RBAC through API or other programmatic means?

When we ask, "What data will this tool expose?" we are getting to the purpose of the tool. For a visualization tool, it is safe to assume that any and all data in Snowflake could be fair game. For a data cataloging tool, it will vary greatly from tool to tool. Some of these may only consume metadata and not any proper data, which, depending on your organization's practices, may not be a concern at all. Likewise, a logging tool may only contain metadata like when tables were loaded from which locations and which roles were granted. In those cases, RBAC may not be a concern.

A similar question to the previous one is "What data will the tool consume?" This question attempts to get at the differences between what data is presented to the end users, and what data the tool processes. The answer to this is more relevant for compliance reasons; confidential data should not be accessible to a tool unless required. For a tool that only needs metadata, we can allocate a service account for the tool that only has privileges on metadata and does not have any access to proper data contained in tables and views. We may even want to create row access policies and masking policies for tools that may need to look at some data but should not surface sensitive data.

When we think about "Who will use this tool?" we want to understand this in the context of the preceding information. We may not be concerned about who is using a tool if the tool does not expose data to the end user. However, if the tool is able to expose proper data to users, then we need to understand who will be using the tool and who will see this data. For data visualization tools, the audience is often the entire organization, or a similarly large group like the business organization. For tools like logging, we might only expect security and data engineers as users of the tool. For a data catalog tool, we might have data scientists, analysts, and developers as our primary users.

The final major question we have to ask is "Does the tool support Snowflake's access control, or does it support RBAC through API or other programmatic means?" which gives us an idea of how we could manage access control in this tool if we determined that we need to based on our answers to the previous questions. We want to understand if we can force users to authenticate through the tool to Snowflake using OAuth, which allows the RBAC system in Snowflake to propagate to the data presented to our users. This is the best option if we are using Snowflake as the source of truth, as it relies on the privileges users have in Snowflake. It is possible that these tools do not support OAuth, but they might support passing user parameters through to Snowflake, which can be invoked in secure views, functions, row access policies, and column masking policies. It also may be that neither of these are supported, but the tool provides an API for groups or roles in the tool. If this is the case, we can leverage an API to create groups or roles based on information we pull from Snowflake. If we don't have any of the preceding options, this presents a security risk as we have created a hole in our RBAC system, allowing users to bypass our protections.

Services That Are Upstream and Downstream

Most systems will also include tools that are both upstream and downstream of Snowflake. An example of this would be tools supporting our SQL developers, data scientists, and other data manipulators. We may have tools like *dbt*, or *MozartData*, which allow the transformation of tables and views in SQL. We can see how this might look in Figure 17-4.

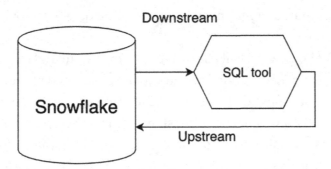

Figure 17-4. *A SQL tool that is both upstream and downstream of Snowflake*

When we work with services that are both upstream and downstream of Snowflake, we're pulling data, transforming it in some way, and inserting the data back into Snowflake. All of the same principles for working with upstream and downstream services apply. We need to understand who will be using these tools, how we can ensure only the correct users have access to view data, and how only the correct users have privileges to edit or update data. Most SaaS tools allow us to lean on the privileges we have created in Snowflake, and if we're using any home-grown tools, we can make sure they respect those boundaries.

Key Takeaways

Maintaining access control across upstream and downstream services of Snowflake is a difficult task, but it is necessary to maintain the integrity of the system. Any point of inconsistency can create an opportunity for data leakage. There are many strategies we can embrace to help maintain consistency, and like with everything else, every organization's needs are different. In the next chapter, we will cover strategies around managing access requests. The following are the key takeaways from this chapter:

- Upstream services include any service, system, or tool that data passes through before being loaded into Snowflake.

- Downstream services include any service, system, or toll that uses data after it has been loaded into Snowflake.

- Downstream services can allow users to authenticate to Snowflake, so data is pulled using their credentials.

- Some tools are both upstream and downstream of Snowflake; these tools will use principles from both.

- RBAC in upstream and downstream services depend on the functionality of tools, so vetting tools is very important.

CHAPTER 18

Managing Access Requests

In the last chapter, we covered how to work with upstream and downstream services that handle Snowflake data. In this final chapter, we're going to explore how we can efficiently manage roles allocated to users. We've covered how to grant read, write, admin, and other types of access to roles, but now we're shifting our focus to how we handle these requests at scale. We're going to go in order of operations, first covering how we can initially create and populate a role. Then we'll cover how a user might be able to discover which roles exist and then finally, we will cover how users can request access to roles.

Role Creation and Population

Before users can request access to roles, we need to create them. In Chapters 5 and 6, we broadly covered how to work with functional and team roles. Now we're going to address creating these roles from the organizational and operational sides. We essentially have a few options for creating these roles, we can manually create them, we can use an employee identity management tool to automatically create roles, or we can use another scripted means using employee data. For all of these methods, we need to solve for the initial bulk creation as well as ongoing creation of new roles. We can also use a combination of these methods.

When we manually create and populate roles, we're having a member of a designated Snowflake team create roles one at a time, and then add users to that role. We don't need a script to generate this SQL, though it never hurts to define SQL in a Python script for easy generation. We simply `CREATE ROLE <ROLE NAME>;` and then `GRANT ROLE <ROLE NAME> TO USER <USERNAME>;` to create and grant a role to a user. We could grant privileges to the role before or after granting the role to users.

© Jessica Megan Larson 2022
J. M. Larson, *Snowflake Access Control*, https://doi.org/10.1007/978-1-4842-8038-6_18

We can also use an SCIM integration to create roles and grant roles to users. This would take advantage of an identity management tool like *Okta* to dynamically create and update groups of users allocated to roles. Ideally, this will match up with existing groups in an organization for consistency, but also for ease of use. One thing to watch out for is when using the Snowflake-specific SCIM connector in Okta, groups cannot be created with more than 100 users. This is a known issue with this connector and the workaround is to initially create groups that are empty, and then add users to those groups afterward, or to use the generic SCIM 2.0 API. Once created, the groups will not have size limitations. Additionally, SCIM can only manage roles created through SCIM. This means that roles created directly in Snowflake, or through the Snowflake Python connector, will not be able to be managed through SCIM.

The last method of creating roles in Snowflake is through using a custom script that pulls configuration data from an employee administration tool or other data source like LDAP, or custom mappings. Once we create the mappings and initial roles, we can have a scheduled task that syncs a list of roles that should exist in Snowflake with the existing roles. This task could add or remove roles and add or remove users from roles.

Role Discovery

Now that we've created roles with access to datasets in Snowflake, and we've granted roles to users, we need to have a way for users to discover additional datasets they may need access to, and a way to map these datasets to roles. We can solve this in two main ways, through a data catalog tool or through careful permissioning, allowing users to see that datasets exist without them being able to see the underlying data until they have access.

Data catalog tools like *Collibra* are helpful in administration of a data warehouse because they provide additional information and context for datasets. These tools provide dataset metadata in addition to things like dataset owners, common users, field definitions, and descriptions of data. This should be enough information for users to make a decision on whether or not they need access to a dataset, and who to reach out to for more information.

When a user requests access to a dataset, the Snowflake administrator can query to see what roles have access to a dataset like

```
SHOW GRANTS ON SCHEMA <SCHEMA NAME>;
```

This query will return all of the privileges set on an object but will not include future grants. If we remember from Chapter 4, future grants are shown using a different query:

```
SHOW FUTURE GRANTS ON SCHEMA <SCHEMA NAME>;
```

This will give us more information about a dataset because it will give us a better idea of read access on a particular schema.

We can also map datasets to roles through a standard naming convention, like having schema `test_schema` map to roles `test_schema_read` and `test_schema_write`, like we covered a few times throughout this book. Another option is to populate the description in the data catalog tool to include the required role(s) or have that information on a wiki page somewhere. The biggest issue with this is that it would likely be manually updated, which means that the process of creating new datasets must include the step of updating this information.

Another way to allow users to discover datasets is through careful permissioning of databases and schemas. If we look back to earlier chapters, we remember that `USAGE` on a database allows a user to see the schemas in a database without being able to directly query the data unless `SELECT` is granted on those objects as well. Additionally, granting users `USAGE` on the contained schemas would allow users to view the tables, views, and other schema objects as well, without being able to directly query either. It is possible that your organization has data that is so sensitive that the metadata itself is sensitive, in this case, a good option is to separate data like that into its own database. To allow users to discover data they do not have access to, I recommend having a role equivalent to `PUBLIC` that is granted to all users, that has `USAGE` on all databases, and `USAGE` on all schemas, without having any object-level privileges.

Requests for Existing Roles

The next step we need to cover now that we've created roles and allowed users to discover datasets, is to field access requests to these datasets. Fielding an access request typically requires a few checks: validating that the user is authorized to access the dataset, logging the request, and granting the request. For datasets under Sarbanes-Oxley (SOX) internal controls, following a defined process for the request, approval, and grant is typically required.

Role Owners

As we covered in Chapters 5 and 6, every role should have a role owner responsible for approving access requests for that role, and modifications to the underlying dataset. An internal wiki page should log this so that it can be referenced any time a request is made.

With Grant Option

Instead of the Snowflake administration team handing access requests for certain or all datasets, we can include WITH GRANT OPTION when we grant privileges to certain roles so that users with those roles can manage grants to their datasets on their own. This means for every privilege on a dataset that may be granted, we would grant the role the privilege like

```
GRANT <PRIVILEGE> ON SCHEMA <SCHEMA NAME> TO ROLE <ROLE NAME> WITH
GRANT OPTION;
```

This allows the initial role to grant that privilege to other roles. We could also do this piecewise, for certain privileges and not others. For example, we could have a role REVENUE_ACCOUNTING_OWNER with most privileges on the REVENUE_ACCOUNTING schema, but we only want them to be able to grant read privileges to other roles. We could do this like

```
GRANT SELECT ON ALL TABLES IN SCHEMA REVENUE_ACCOUNTING TO ROLE REVENUE_
ACCOUNTING_OWNER WITH GRANT OPTION;
GRANT CREATE TABLE ON SCHEMA REVENUE_ACCOUNTING TO ROLE REVENUE_
ACCOUNTING_OWNER;
```

This means that a user with REVENUE_ACCOUNTING_OWNER could successfully run

```
GRANT SELECT ON ALL TABLES IN SCHEMA REVENUE_ACCOUNTING TO ROLE ACCOUNTING;
```

But this user would not be able to run

```
GRANT CREATE TABLE ON SCHEMA REVENUE_ACCOUNTING TO ROLE ACCOUNTING;
```

The preceding snippet would fail because the REVENUE_ACCOUNTING_OWNER role does not have the grant option on CREATE TABLE.

One big drawback from using this method of fielding access requests is that it is not consistent. It is unlikely that roles will have consistent access for reading, writing, and administering datasets. For this reason, I recommend using this approach for sandbox style development datasets. This allows leads, managers, and developers to quickly collaborate on development, without creating inconsistent access on production datasets.

Using a Ticketing System

For managed roles, using a ticketing system like *Jira* can help manage incoming requests while at the same time logging these requests for potential future auditing. The process will likely look like this:

1.   User creates a ticket asking for access to a particular dataset.

2.   Snowflake administrator tags the appropriate dataset owner on ticket for approval.

3.   Dataset owner approves access request.

4.   Snowflake administrator grants request and closes out the ticket.

Datasets subject to SOX internal controls may require a process similar to this so that the entire flow can be logged. This also creates a paper trail for future reference that can help users request the same permissions that a teammate requested previously.

Building a Custom Tool

Another option is to create a custom tool using Snowflake's Python connector where users can request access to datasets. This tool could allow users to browse datasets and select which datasets they need access to, creating an access request. Dataset owners could then field these requests, which triggers the application backend to connect to Snowflake using a service account to grant access automatically. A solution like this would require some heavy lifting, but may be worth it for your organization if it is sufficiently large and an automated tool is desired.

Key Takeaways

Managing access requests is a task that will almost certainly be here to stay, but there are ways we can minimize the lift required while at the same time maintaining the security of our system. By using a systematic approach and some engineering resources, we can scale the management of access requests. The following are key takeaways from this chapter:

- Roles can be created and populated manually, using an identity management tool, or through custom code with role configuration.

- A data catalog tool can help users browse datasets they may need access to in the future, enabling them to request access.

- Using the `WITH GRANT OPTION` allows a role to grant privileges to other roles, which can allow teams to work more quickly.

- Ticketing systems can help Snowflake administration teams manage access requests and comply with regulations like SOX.

- Organizations can also build custom tools to manage access requests, allowing users to browse datasets and request access.

Index

A

Access control, 3, 8
 methods, 6
  groups/roles, 7
  lookup tables/mappings, 7
 paradigms, 4
  attribute-based access
   control, 6
  data democratization, 5
  PLP, 5
  RBAC, 5
  rule-based access control, 6
  separation of duties, 6
Access request
 request for existing roles, 235
  custom tool, 237
  grant option, 236, 237
  role owners, 236
  ticketing system, 237
 role creation/population, 233, 234
 role discovery, 234, 235
Account-level privileges, 126
 account-level objects, 125
 monitoring activity, 125
 user/role management, 124
Account object privileges, 45
Act on the Protection of Personal
  Information (APPI), 30, 31
 definition, 31
 handling data, 31
Administrative privileges, 131, 133
Admin privileges, 116, 145

Aggregated data, 19
Aggregate statistics, 168, 169
Aggregation, 19
ALTER commands, 145
ALTER SHARE command, 199
American Civil Liberties Union
  (ACLU), 26
Analytics environment, 224
Anonymized data, 15–19, 34
Apache Airflow, 227
Application Programming Interface (API),
  138, 187, 226, 229, 234
Attribute-based access control, 6, 7

B

Big data, 25
Boolean SQL statements, 156–158

C

California Consumer Privacy Act (CCPA),
  9, 31, 32
 definition, 33
 handling data, 33
California Consumer Protection Act
  (CCPA), 3
California Privacy Rights Act of 2020
  (CPRA), 32, 33
Centralized data team, 71, 75, 77, 134
Cloning, 213, 215, 216
Cloud storage, 213, 214, 225, 227
Cloud storage platform, 201

© Jessica Megan Larson 2022
J. M. Larson, *Snowflake Access Control*, https://doi.org/10.1007/978-1-4842-8038-6